AUDEL®

Plumbers and Pipe Fitters Library Volume I

Materials
Tools
Roughing-In

by CHARLES McCONNELL

Macmillan Publishing Company
New York

Collier Macmillan Publishers
London

FOURTH EDITION

Macmillan Publishing Company
866 Third Avenue, New York, NY 10022
Collier Macmillan Canada, Inc.

Library of Congress Cataloging-in-Publication Data
McConnell, Charles.
  Plumbers and pipe fitters library.
  Includes indexes.
  Contents: v. 1. Materials, tools, roughing-in—
v. 2. Welding, heating, air conditioning—v. 3. Water
supply, drainage, calculations.
  1. Plumbing—Handbooks, manuals, etc.   2. Pipe-
fitting—Handbooks, manuals, etc.   I. Title.
TH6125.M4   1988   696'.1   88-8815

ISBN 0-02-582911-4 (v. 1)
ISBN 0-02-582914-9 (set)

Macmillan books are available at special discounts for bulk purchases
for sales promotions, premiums, fund-raising, or educational use.
For details, contact:

        Special Sales Director
        Macmillan Publishing Company
        866 Third Avenue
        New York, NY 10022

10 9 8 7 6 5 4 3 2 1

Printed in the United States of America

Foreword

Plumbing and pipe fitting play a major role in the construction of every residential, commercial, and industrial building. Of all the building trades, none are as essential to the health and well-being of the community as the plumbing trade. In addition, the knowledge and craftsmanship of the skilled pipe fitter is required in shipfitting, aircraft, and space vehicles.

Pollution of lakes, rivers, and the aquifer is of vital concern to everyone and is addressed in this three-volume series. Modern technology has created new materials which have revolutionized the piping trades. Materials such as lead and cast iron, used for decades, are being supplanted by plastics.

Plumbing and pipe-fitting installations are governed by codes, regulations, and ordinances established by local, state, and federal agencies. Inspections by licensed inspectors are designed to insure that all rules and regulations governing the work are complied with. Plumbers, and in many areas pipe fitters, are required to pass examinations and be licensed in order to engage in the trade. Health and safety of the population and environmental and ecological concerns demand highly skilled workmen in the building trades. This three-volume series has been written to provide a reference source for those already engaged in the plumbing and pipe-fitting trades and as an aid to those whose intention it is to become a plumber or pipe fitter.

This, the first of three volumes, deals with materials used by the piping trades, pipe fittings and valves, steel piping, blueprints and elevations, cast-iron soil pipe, roughing-in plumbing, plumbing fixtures, and lead work.

Contents

Acknowledgments

The author wishes to thank the following companies for their assistance in furnishing information and drawings on their products:

Asahi/America, Inc.
Cast-Iron Soil Pipe Institute
Crane Plumbing
David White Instruments
Delta Faucet Co.
Eljer Plumbingware
Fluidmaster, Inc.
Hayward Industrial Products, Inc.

L. S. Starrett Co.
Moen/Stanadyne Group
Owens-Corning Fiberglas
The Powers Regulator Co.
Ridge Tool Co.
Sloan Valve Co.
Stanley Tools
Watts Regulator Co.

CHAPTER I

Materials Used by the Piping Trades

Plumbing and pipe fitting began in ancient Rome; and the word *plumbing* is derived from the Latin *plumbum,* the art of working with lead. For centuries the principal materials used by plumbers and pipe fitters were lead and cast iron; ductile iron, steel, brass, copper, and glass came into use comparatively recently. Plastics, superior in many respects to each of the above, are rapidly superseding these materials.

Knowledge of the composition and uses of these materials is essential for the proper installation of plumbing and its related trade, pipe fitting. In this chapter we will explore the composition and uses of the above-mentioned materials and some others. In other chapters of this three-volume series we will explain their uses and methods of application and installation in greater depth.

Lead

Lead was a much-used plumbing material until the 1960's when, due to the increasing use of plastics, it began to be replaced by other materials. Lead is the heaviest of all common metals, weighing .4106 lbs. per sq. in. It is a bluish gray metal with a bright luster when melted or newly cut. Commercial lead has a lower specific gravity than pure lead (11.37) because of the impurities it contains.

The safe working strength of lead is about one-fourth of its elastic limit, or 225 lbs. per sq. in. It is very soft, especially when allowed to cool and solidify slowly. Lead does not crystallize readily. When refined, lead is poured at the correct temperature into a warm mold and allowed to cool. Fernlike crystalline aggregates appear at the surface. In the form of filings, lead becomes a solid mass if subjected to a pressure of 13 tons per sq. in. and liquefies at 2½ times this pressure. Lead undergoes no change in dry air or in water that is free from air. It becomes pasty at about 617°F and melts at about 650°F. It boils at about 15,000°C but cannot be distilled; its coefficient of linear expansion at ordinary temperatures is .00001571 per 1°F. The strength of lead in both compression and tension is very small. Since lead unites readily with almost all other metals, it is used in many alloys for bearing metals, solders, and the like. Alloys composed of lead, bismuth, and tin are noted for their low melting points.

Sheet lead is used for lining rooms containing X-ray equipment, and to shield and protect the X-ray technician from exposure to the rays. Application of the sheet lead requires great skill in "burning" (welding) the joints in sheet lead.

Lead poisoning can occur both from breathing the fumes from melting lead and by handling lead. The following precautions should be taken when working with this metal to guard against poisoning:

1. Wear a filter type mask when melting lead or when in an area where lead is being melted.
2. Wear protective clothing, overalls, etc., while working with lead. This outside clothing should be washed as frequently as possible.
3. Eat a substantial meal before going to work. With an empty stomach, conditions are more favorable for absorption of lead by the body.
4. Drink water and milk plentifully.
5. Bathe frequently.
6. If you feel at all sick, consult a doctor at once.

Cast Iron

Cast iron is iron containing so much carbon that it is not malleable at any temperature. It consists of a mixture of iron and carbon, with other substances in varying proportions. Generally, commercial cast iron contains between 3 and 4 percent carbon. The carbon may be present as graphite, as in gray cast iron, or in the form of combined carbon, as in white cast iron. In most cases the carbon is present in both forms. Besides carbon, a combination of silica, sulfur, manganese, and phosphorus is nearly always present in some degree.

Cast iron is used for soil, waste and vent piping and fittings, for water-main piping and fittings, natural and manufactured gas mains, steam and hot water heating fittings, valves, cast-iron boiler sections, and many specialty items.

Ductile Cast Iron

Nodular or ductile cast iron is widely used for water-main piping. Standard cast-iron water main has the advantage of being easily cut to length using a heavy hammer and cold-cut type chisel, but due to the rigidity and non-flexing of cast iron, it will break easily if subjected to stress. Ductile cast iron will give slightly under stress, eliminating most problems of broken mains, but due to the composition of ductile iron it cannot be cut to length using the hammer and chisel method. It must be cut using a saw with a carborundum disc blade.

Steel

Steel is an important construction material. Its great strength permits its application to the largest and most severely strained constructive members. It can be forged or cast in any convenient form and is readily obtained in the form of plates, bars, pipe fittings, steel boilers, water heaters, appliances, and plumbing fixtures. A disadvantage is that it is vulnerable to rust and corrosion, requiring systematic and careful attention in order to preserve it against the action of moisture, oxygen, and carbonic acid.

Upon immersion in a polarizing fluid it is also attacked by galvanic action (electrolysis) in connection with copper or brass. In regard to its percentage of carbon, steel occupies a middle position between cast iron and wrought iron. In common with the former, it has a sufficiently low melting point for casting and, in common with the latter, a sufficient toughness for forging. According to their varying percentages of carbon, three kinds of steel may be recognized:

1. Soft steel.
2. Medium steel.
3. Hard steel.

Soft steel is nearest to wrought iron in carbon percentage and qualities; it is soft, readily forged, and by careful handling may also be welded. It is principally used in flanged parts, furnace plates, rivets, water heater tanks, steel boilers, and other materials exposed to alternate heating and cooling or to severe treatment by shaping and forming.

Medium steel is harder than soft steel, and one of its principal uses for pipe fitters is boiler shells.

Cast steel has about the same percentage of carbon as soft or medium steel. In addition it contains silicon and manganese, which are needed to produce good castings. Hard steel comes the nearest to cast iron in carbon percentage and possesses as its most important quality a decided facility for tempering and hardening when cooled quickly in water.

Copper

Copper is a common metal of a brownish red color, both ductile and malleable and very tenacious. It is one of the best conductors of heat and electricity. It is one of the most useful metals in itself and in its various alloys, such as brass and bronze. It is one of many metals that occur native. It is also found in various ores, of which the most important are chalcopyrite, chalcocite, cuprite, and malachite. Mixed with tin, it forms bell metal; with a smaller proportion, bronze; and with zinc, it forms brass, pinchbeck, and other alloys.

The strength of copper decreases rapidly with a rise in temperature above 400°F($\pm$); between 800°F and 900°F, its strength is reduced to about half that at ordinary temperatures. Copper is not easily welded, but may be readily brazed. At near the melting point it oxidizes (or is burned) and loses much of its strength, becoming brittle when cool.

Brass

Brass is a yellow alloy composed of copper and zinc in various proportions. Brass pipe, brass fittings, faucets, and valves are used in many applications by the plumbing and piping trades. When zinc is present in small percentages the color of brass is nearly red; ordinary brass pipe contains from 30 to 40 percent zinc. Brass can be readily cast, rolled into sheets, or drawn into tubes, rods, and wire of small diameter. The composition of brass is determined approximately by its color; red brass contains zinc, 5%; bronze, 10%; light orange, 15%; greenish orange, 20%; yellow, 30%; yellowish white, 60%.

Tin

Tin is a soft white metal with a tinge of yellow. It has a high luster, hence is frequently used as reflectors of light. Tin, when nearly pure, has a specific gravity of 7.28 to 7.4, the pure tin being the lightest. It has a low tenacity but is very malleable and can be rolled or laminated into very thin sheets, known as tinfoil. The melting point of tin is 433°F. At 212°F (the boiling point of water) it is ductile and easily drawn into wire. It boils at white heat. It burns with a brilliant white light when raised to a high temperature and exposed to the air. Its specific heat is .0562; latent heat of fusion, 25.65 btu per lb.

Conductivity is low, and it oxidizes slowly in the air at ordinary temperature. When exposed to extreme cold, tin becomes crystalline. Heat conductivity is 14.5; electric conductivity is 12.4 as compared with silver, which is 100. Its weight is 459 lbs. per cu. ft. Tensile strength is 3500 lbs. per sq. in.; crushing load (cast tin) is 15,500 lbs. per sq. in. Due to its high power of resistance to tarnishing by exposure to air and moisture, tin is used as a protective coating for iron and copper. Diluted sulfuric acid has no action on tin when cold, but when tin is boiled in concentrated acid, the metal is dissolved. Coefficient of expansion of tin is .0000151 per 1°F. The principal use of tin by plumbers is for alloying with lead to make solders.

Glass Pipe

Glass pipe is widely used in a great variety of industries. Dairy products, food, chemical, pharmaceutical processors, and paper and pulp find that the corrosion resistance and transparency of glass pipe make it suitable for applications where observation of the processing is vital. Glass piping in short lengths can be inserted into nontransparent piping for use as sight-flow indicators. Glass pipe is also lightweight and the extra smooth inside surface reduces pressure drop due to friction and discourages scale buildup. Matching glass fittings are made for use with the glass pipe, and several different types of flanged connections and gaskets can be selected to fit the particular job need. Glass pipe can be easily cut to length and can be used with plain ends or beaded on the job if the application makes beading desirable.

Another form of glass piping used widely, especially for acid wastes or acid conducting pipe, is a fiberglass pipe. This pipe is made of fiberglass, spun and coated with a resin; it is tough, lightweight, and impervious to most acids.

Malleable Iron

The method of producing malleable iron is to convert the combined carbon of white cast iron into an amorphous uncombined condition by heating the white cast iron to a temperature somewhere between 1380° and 2000°F. The iron (sometimes called castings) is packed in retorts or annealing pots, together with an oxide of iron (usually hematite ore). The oxygen in the ore absorbs the carbon in the iron, giving the latter a steel-like nature.

An annealing furnace or oven is used for heating, and the castings are kept red hot for several days or several weeks, depending on the pieces. In order for the process to be successful, the iron must have nearly all the carbon in the combined state and must be low in sulfur. Usually, only good charcoal-melted iron that is low in sulfur is used, although a coke-melted iron is suitable, provided the proportion of sulfur is small. The process is not adapted to very large castings because they cool slowly and usually show a considerable proportion of graphite. Malleable iron pipe fittings are one use of this metal.

Oakum

Oakum consists of shredded rope or hemp fiber. It is sold as a dry oakum for packing poured joints in water mains and as tarred or oiled oakum for use in poured joints of soil pipe. White oakum is a fibrous material covered with a thin woven coating and impregnated with a cementlike powder. White oakum swells when brought into contact with water and is excellent for use as a packing material when making soil-pipe joints.

Asphaltum

The name asphaltum is given to a waterproofing paint made from asphalt. Asphalt is black or dark brown in color and will melt or burn, leaving little residue. It dissolves in petroleum or turpentine. It is used for coating pipes and other metals exposed to dampness and weather.

Plastics

Plastics used for pipe and fittings include a number of materials which differ significantly in their properties, characteristics, and suitability for specific jobs. These differences are important in avoiding misapplications. Plastic materials are generally classified in two basic groups, *thermoplastics* and *thermosetting* resins. Thermoplastics can be re-formed repeatedly by applications of heat. Thermosetting resins, once their shape is fixed and cured, cannot be changed for re-use.

PVC (Polyvinyl Chloride), a Thermoplastic

Polymerized vinyl chloride is a synthetic resin which when plasticized or softened with other chemicals has some rubberlike qualities. It is derived from acetylene and anhydrous hydrochloric acid.

PVC is inert to attack by many strong acids, alkalies, salt solutions, alcohols, and many other chemicals. It imparts no taste or odors to materials handled. PVC provides both high tensile and impact strength. It is resistant to fungi, bacterial action, and adverse soil conditions. Thermoplastic pipe and fittings are immune to the electrolytic or galvanic action which attacks metal piping. PVC is light in weight; a 20 ft. length of 4 in. Schedule 40 (standard weight) PVC pipe weighs 40 lbs. The same length and size of standard weight steel pipe weighs 250 lbs.

The inside surfaces of PVC pipe and fittings are mirror smooth, which provides low head and friction loss and higher flow rates. The most commonly used method of joining PVC pipe, fittings, and valves is by solvent welding. Although there are many applications where PVC pipe, fittings, and valves are used, plumbers and pipe fitters work with this material primarily in the installation of soil, waste, vent, and process piping.

CPVC (Chlorinated Polyvinyl Chloride)

The CPVC molecule has one more chlorine atom than the PVC molecule. This extra chlorine is responsible for the material's high-temperature strength and other properties so valuable for industrial piping.

CPVC has an upper temperature working limit of 180°F, approximately 40° above that of other rigid vinyl thermoplastics. For pressure piping applications it is recommended for temperatures as high as 180°F.

The installed cost of CPVC piping is considerably below that of expensive alloys. CPVC is an economical material for process piping, hot water lines, and similar applications where operating conditions exceed the recommended temperature limits for PVC. CPVC has a tensile strength of more than 2000 psi at 212°F.

Cementing and Welding

One method of joining thermoplastics is to use a solvent as an adhesive. The socket of the fitting and the male end of the pipe should first be cleaned, using a PVC cleaner or acetone. The solvent, PVC cement, should be applied immediately to both the socket of the fitting and the male end of the pipe; the pipe should be inserted into the fitting, lined up correctly, and held in place for approximately 30 seconds. The solvent softens the molecules on the surface of both the pipe and the fitting and when the two surfaces are pressed together the molecules fuse tightly, forming a very strong bond. Solvent welding, if done correctly, provides a strong and waterproof joint and is the favored method of joining PVC soil, waste and vent piping, and CPVC domestic water piping.

Thermoplastic materials may also be softened and joined by heating, which permits their molecules to move relative to one another and so enables a joint to be made when the softened ends are brought together.

Detailed information on the use of plastic piping will be found in the chapter on Process Piping in Volume III of PLUMBERS AND PIPE FITTERS LIBRARY.

Crosslinked Polyethylene (CAB-XL)

CAB-XL is a material that has excellent strength characteristics and improved resistance to most chemicals and solvents at elevated temperatures to 203°F. Crosslinking permits high impact strength even at subzero temperatures. This material is often suggested for services too severe for ordinary polyethylene.

Glass-Reinforced Epoxy

Glass-reinforced epoxy is probably the best thermoset plastic for piping applications. It has a high strength-to-weight ratio and an outstanding resistance to chemicals and weathering.

Polypropylene

Polypropylene is the lightest of the thermoplastic piping materials, yet has higher strength and better general chemical resistance than polyethylene and may be used at temperatures 30° to 50° above the recommended limits of polyethylene. Polypropylene is an excellent material for laboratory and industrial drainage pipe where mixtures of acids and solvents are involved. It has found wide application in the petroleum industry where its resistance to sulfur-bearing compounds is particularly useful in salt-water disposal lines, low-pressure gas-gathering systems, and crude-oil flow piping. It is best joined by fusion welding.

Polyethylene

Polyethylene is the least expensive of the thermoplastics and one of the most widely used. Although its mechanical strength is comparatively low, it exhibits very good chemical resistance and is generally satisfactory when used at temperatures below 120°F. Types I and II (low- and medium-density) are used frequently in chemical laboratory drainage lines, field irrigation, and portable water systems. Fusion welding is the best method for joining this material.

Acrylonitrile-Butadiene-Styrene (ABS)

ABS has high impact strength, is very tough, and may be used at temperatures up to 180°F. It has a lower chemical resistance and lower design strength than PVC. ABS is used for carrying water for irrigation, gas transmission, drain lines, waste and vent piping. Solvent welding and threading are recommended fabrication methods.

CHAPTER 2

Basic Tools

The plumber and pipe fitter are expected to know how to use a greater variety of tools than any other building tradesman. Many specialized tools are needed. The tools listed and illustrated in this chapter are common. Some specialized tools will be pictured and described in chapters describing the work in which they are used.

Rules

A good six-foot rule is a necessary tool for the plumber and pipe fitter. A wooden folding *inside reading* rule (Fig. 2-1) is the most practical rule for several reasons. It can be opened partway, say to the 30-in. mark, and when laid on a set of plans, the rule will lay flat, thus showing more accurate measurements. The numbers from 1 to 6 are used more often than 66 to 72, and the low numbers, being on the inside, are protected from wear.

Measuring Tapes

Measuring tapes can be purchased in a variety of lengths and types, but the most common are the 50- and 100- ft. lengths. For measuring

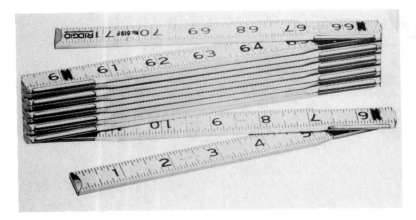

Fig. 2-1.   A wooden folding rule. *(Courtesy Ridge Tool Co.)*

distances of 6 to 10 ft., a pocket type with a pushbutton spring wind is available. The longer tapes, as shown in Fig. 2-2, are normally used for measuring pipe for installations requiring long runs. 50- and 100-ft. measurements can be made accurately using the type of steel tape shown in Fig. 2-2.

Fig. 2-2.   A 50-foot steel measuring tape. *(Courtesy L.S. Starrett Co.)*

Fig. 2-3. **A scratch awl.**

Marking, or Scratch, Awls

An awl consists of a short piece of round steel, pointed at one end and with the other end fixed in a knob handle. An awl is used as a scriber and as a center punch. It is also useful for lining up holes in companion flanges, etc. A scratch awl is shown in Fig. 2-3.

Levels

A level is used to check horizontal objects or material; pipe, bases, equipment, for true level; or for a required pitch or slope. It is also used to insure that perpendicular objects or material are truly vertical or at 90° relative to a level base.

A level can be tested for accuracy by placing it on or against a surface and marking the surface at each end of the level. Observe the position of the bubble, then reverse the level, turning it 180° and place it within the marked lines. If the bubble is not in exactly the same position as it was before it was reversed, the level is faulty. A level is shown in fig. 2-4.

Fig. 2-4. **A level.** *(Courtesy Ridge Tool Co.)*

Chisels and Punches

The plumber frequently must use the chisel to cut cast-iron pipe. There are several variations of chisel for specific jobs (Fig. 2-5).

(A) Cold chisel used to cut slag from welds, cut cast-iron soil pipe, cut concrete, etc.

(B) Diamond point chisel, all-purpose, for removing slag from welds, cutting out broken threads, etc.

(C) Prick punch, for laying out patterns, marking cuts.

(D) Center punch, used primarily to indent a surface for starting a drill bit.

(E) Drift pin, used to align holes in valves and flanges.

Fig. 2-5. Chisels and punches used by plumbers and pipe fitters. *(Courtesy Stanley Tools)*

Plumb Bobs

The word *plumb* means *perpendicular to the plane of the horizon,*
and since the plane of the horizon is perpendicular to the direction
of gravity at any given point, the force due to gravity is utilized to
obtain a vertical line in the device known as a plumb bob, shown
in Fig. 2-6. The plumb bob is made from solid steel, bored and
filled with mercury to provide a low center of gravity and great
weight in proportion to its short length and small diameter. A no-
roll hex head prevents rolling when the plumb bob is set down.
Since the point is removable, it can be easily replaced if broken or
worn. An outstanding feature that results in the bob hanging perfectly true is the device for fastening the string, without a knot to
tie or untie, by simply drawing it into the specially shaped slotted
neck at the top.

Fig. 2-6.   A plumb bob.
(Courtesy L.S. Starrett Co.)

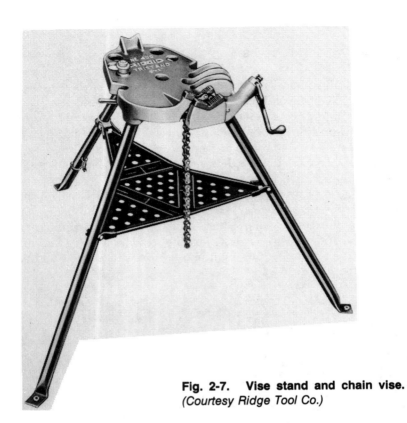

Fig. 2-7. Vise stand and chain vise.
(Courtesy Ridge Tool Co.)

Fig. 2-8. A yoke vise.
(Courtesy Ridge Tool Co.)

Vise Stands and Vises

Vise stands and vises are used to hold pipe for cutting, threading, and tightening or removing fittings. Either yoke vises or chain vises can be mounted on the stand shown in Fig. 2-7. The chain vise shown in Fig. 2-7 will hold ¼-in. through 6-in. pipe. The yoke vise shown in Fig. 2-8 will hold ⅛-in. through 2½-in. pipe.

Pipe Cutters

Figure 2-9 shows a typical pipe cutter that can be used by hand or with power equipment. It is adjustable to cut pipe from ⅛ in. through 2 in.

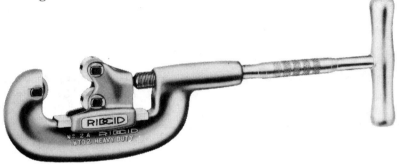

Fig. 2-9. This wheel and roller pipe cutter will cut ⅛″ I.D. through 2″ I.D. pipe. *(Courtesy Ridge Tool Co.)*

Fig. 2-10. A tubing cutter for 2⅛″ I.D. through 4″ I.D. tubing. *(Courtesy Ridge Tool Co.)*

Fig. 2-11. A hinged pipe cutter. Tubing cutters, in various sizes, can be used to cut copper tubing from ⅛" I.D. through 4" I.D. *(Courtesy Ridge Tool Co.)*

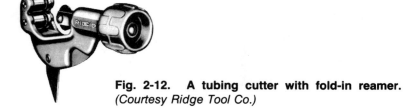

Fig. 2-12. A tubing cutter with fold-in reamer. *(Courtesy Ridge Tool Co.)*

Fig. 2-13. A midget tubing cutter. *(Courtesy Ridge Tool Co.)*

Midget tubing cutters, Fig. 2-13, are designed for use in extra-close quarters on small size (up to ¾-in. I.D.) copper or aluminum tubing.

Soil Pipe Cutters

Tools have been developed that make a hammer and chisel method of cutting soil pipe obsolete. Also, the trend in the industry is to change from extra-heavy soil pipe to the new, light-service-weight soil pipe. For cutting extra-heavy soil pipe, clay tile, cement pipe, and Class 22 water main, a chain cutter such as is shown in Fig. 2-14A is used; for 2- through 6-in. NO-HUB soil pipe, a cutter such as is shown in Fig. 2-14B is best, due to the closer spacing of the cutter wheels. The operation of both types of cutter is the same; the jaws are opened by turning the adjusting knob, the chain is locked around the pipe and placed in the jaw notch, the ratchet knob is set with the arrow pointing down, and a few easy pumps tighten the chain until the pipe is severed.

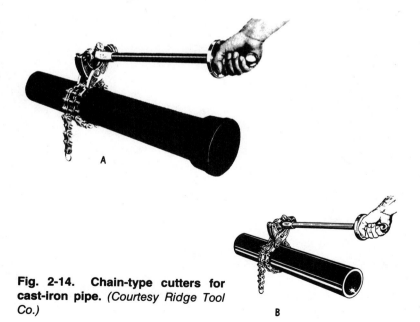

A

B

Fig. 2-14. Chain-type cutters for cast-iron pipe. *(Courtesy Ridge Tool Co.)*

Fig. 2-15. **A plastic tubing cutter.** *(Courtesy Ridge Tool Co.)*

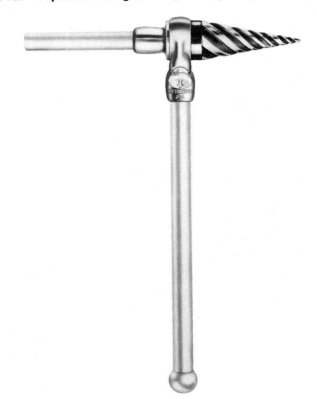

Fig. 2-16. **A self-feeding reamer for ⅛″ I.D. through 2″ I.D. pipe.** *(Courtesy Ridge Tool Co.)*

Pipe Reamers

Pipe cut or threaded by any method should be reamed to remove inside burrs. The spiral reamer in Fig. 2-16 is self-feeding and can be used by hand or with power equipment.

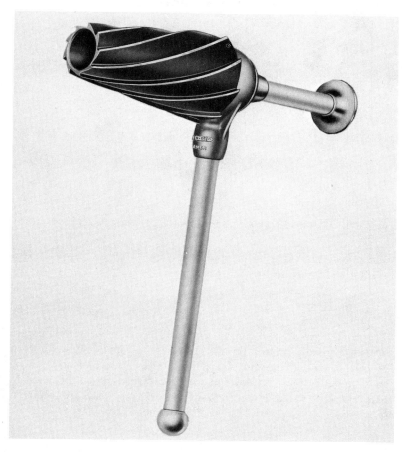

Fig. 2-17.   A self-feeding reamer for 2½" I.D. through 4" I.D. pipe. *(Courtesy Ridge Tool Co.)*

Pipe Threaders

There are two models of the stock and die set shown in Fig. 2-18. One will thread ⅜-in., ½-in., and ¾-in. pipe, the larger set can be

Fig. 2-18.   A three-way stock and dies. *(Courtesy Ridge Tool Co.)*

used to thread ½-in., ¾-in., and 1-in. pipe. The die segments can be reversed for close-to-wall threading.

⅛ to 1 Inch

The ratchet threader shown in Fig. 2-19 is available with separate heads to thread pipe from ⅛-in. through 1-in. Ratchet-type dies can be used to thread pipe in place, if necessary, and can be used by hand or with power equipment. Similar threaders can be used to thread pipe up to and including 2-in. in size. The heads are removed from the stock by pulling out on the ratchet knob.

1 to 2 Inches

The adjustable ratchet threader shown in Fig. 2-20 uses one set of dies to thread pipe from 1-in. up to and including 2-in. Using this type threader, the depth of thread is also adjustable, by varying the starting point of the threading head. There are two steps to threading pipe with this threader. First, the dies and the rear chuck must be

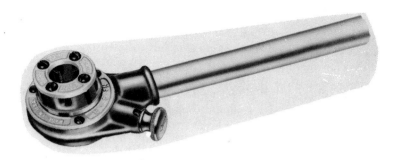

Fig. 2-19. A ⅛″ to 1″ ratchet threader. *(Courtesy Ridge Tool Co.)*

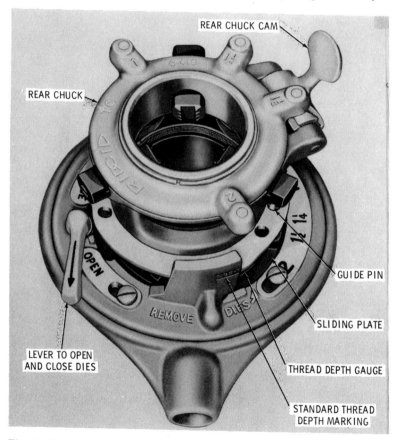

Fig. 2-20. A 1″ to 2″ ratchet threader. *(Courtesy Ridge Tool Co.)*

set for the pipe size. The rear chuck face should be turned to the pipe-size setting and the locking cam left in open position. To set the dies for the pipe size, turn the rear chuck counterclockwise until the pins, which set the die head, are free from the plate. When free, the pins can be moved to line up with the proper pipe-size marking. The rear chuck should then be turned clockwise, and the pins will enter the alignment holes in the plate. Turn the rear chuck until the plate is in line with the *STD* (standard thread depth marking) on the threader body. Slide the threader onto the pipe, center the pipe against the pipe dies, and lock the cam on the rear chuck. The dies are now ready to cut a standard depth thread. The starting points for deeper or shallower than normal threads are shown in Fig. 2-20.

Fig. 2-21. **A 2½″ to 4″ geared threader.** *(Courtesy Ridge Tool Co.)*

2½ to 4 Inches

The geared adjustable pipe threader shown in Fig. 2-21 uses one set of dies to thread 2½-, 3-, 3½-, and 4-in. pipe. It can be used for hand threading or with a power machine, and the depth of thread can be varied to suit a given condition.

Portable Power Drives

Portable power drives, such as the one shown in Fig. 2-22, are used to hold pipe from ⅛-in. I.D. through 2-in. I.D. for threading. The power drive can be used to thread pipe larger than 2-in. I.D. by using a drive shaft with universal joints coupled to the threader or by using close-coupled drive accessories for geared threaders.

Fig. 2-22.   A portable power drive. *(Courtesy Ridge Tool Co.)*

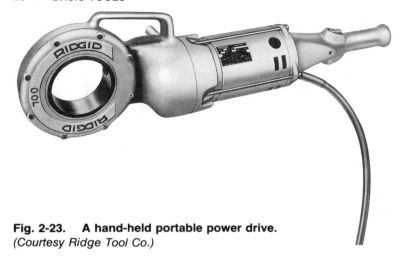

Fig. 2-23. A hand-held portable power drive.
(Courtesy Ridge Tool Co.)

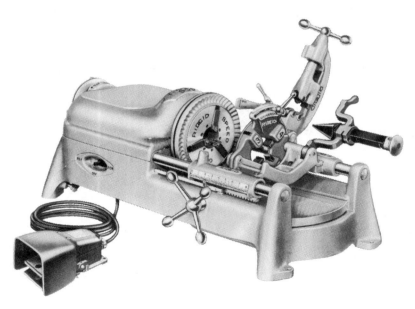

Fig. 2-24. ⅛" to 2" pipe and rod threading machine. *(Courtesy Ridge Tool Co.)*

The portable power drive shown in Fig. 2-23 can be used to drive the threader shown in Fig. 2-22 and to drive hoists, operate large valves, and many other applications. The power drive is reversible.

Pipe and Rod Threading Machines

Pipe machines are used to save both time and labor. The pipe machine shown in Fig. 2-24 will thread pipe from ⅛ in. through 2 in. and rod from ¼ in. through 2 in. The die heads used are the quick-opening type; the dies, cutter, reamer, and oiler swing out of the way when not in use. The machine can be mounted on a wheeled stand for easy moving.

2½- through 4-in. pipe can also be cut and threaded on a pipe machine. The machine shown in Fig. 2-25 uses quick-opening die

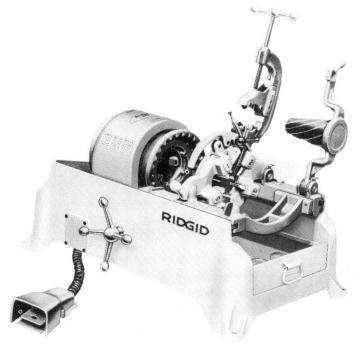

Fig. 2-25. **2½″ to 4″ pipe threading machine.** *(Courtesy Ridge Tool Co.)*

heads; the reamer, cutter, and oiler swing out of the way when not in use. This machine can either be bench mounted or mounted on a wheeled stand for easy moving.

Pipe-Bending Tools

There are a variety of devices for bending pipe, both power and hand-operated. One of the most common pipe benders is illustrated in Fig. 2-26. This lever-type bender is versatile, accurate, and easy to use. Correctly measured, bends will be accurate to blueprint dimensions within plus-minus ⅟₃₂ in. A scale on the link eliminates extra measuring and assures fast tube positioning for accurate finished dimensions. To eliminate possible hand injury, handles are held wide apart when making 180° bends. There are six sizes for soft or hard copper, brass, aluminum, steel, and stainless-steel tube.

Fig. 2-26.   Lever-type tube bender.

Flaring Tools

Several different flaring tools are often needed by the plumber and pipe fitter. The type shown in Fig. 2-27A is called a flaring block and will flare several different sizes of tubing. It is made as a unit; the yoke cannot come off and be lost. The feed releases automatically when the flare is completed.

The flaring tool shown in Fig. 2-27B is a hammer-type tool and is usually used for flaring copper water tubing.

Slip-joint, or adjustable, pliers are an all-purpose tool for plumbers and pipe fitters. Useful for gripping, tightening, or loosening nuts, bolts, packing glands, bonnets, etc.

A

B

Fig. 2-27. Typical flaring tools.
(Courtesy Ridge Tool Co.)

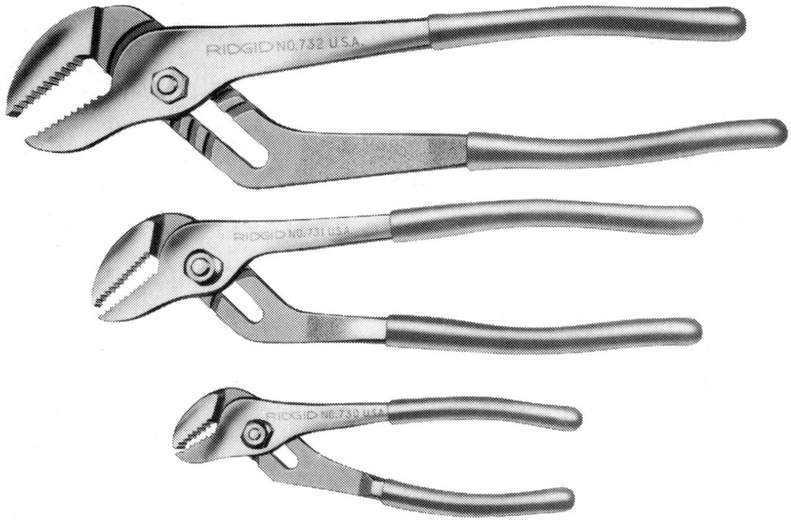

Fig. 2-28. Slip-joint pliers. *(Courtesy Ridge Tool Co.)*

Fig. 2-29. A straight-pattern pipe wrench. *(Courtesy Ridge Tool Co.)*

Fig. 2-30. An end-pattern pipe wrench. *(Courtesy Ridge Tool Co.)*

Pipe Wrenches

The *straight-pattern pipe wrench* (Fig. 2-29), available in six sizes from 6 in. through 60 in., is the basic tool of this trade. It is used to install and remove pipe and fittings.

The *end-pattern pipe wrench* (Fig. 2-30) is used when working in close quarters or next to a wall or corner. Available in eight sizes, 6 in. through 36 in.

The *chain wrench* (Fig. 2-31), which can be used with a ratchetlike action, makes it easier to work in extra close quarters. Made in four sizes from 14 in. through 36 in.

The *compound wrench* (Fig. 2-32) lets one man do the work of two. The short handle makes it easy to get at frozen joints in tight quarters. The turning force of the wrench is multiplied by compound leverage.

Pipe Taps

Pipe taps are used primarily to re-cut damaged threads in fittings or in equipment. Pipe *taps* are used to cut *internal* threads whereas

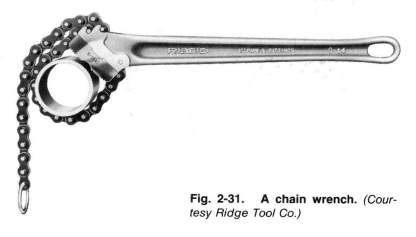

Fig. 2-31. A chain wrench. *(Courtesy Ridge Tool Co.)*

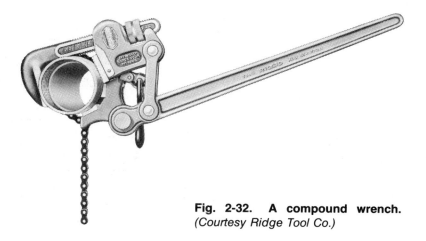

Fig. 2-32. A compound wrench. *(Courtesy Ridge Tool Co.)*

pipe *dies* cut *external* threads. Pipe taps can be used to cut new threads in metal plating or other equipment. Thread-cutting oil should be applied liberally when using a pipe tap and care must be taken to start the tap straight. A typical pipe tap is shown in Fig. 2-33.

When a new thread is to be tapped, the correct size hole for the tap must be drilled. Table 2-1 shows the correct size drill to use for various size taps.

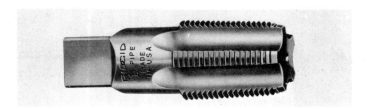

Fig. 2-33.   A pipe tap for cutting internal (female) threads. *(Courtesy Ridge Tool Co.)*

Table 2-1   Tap and Drill Sizes

Based on Approximately 75% Full Thread

National Coarse & Fine Threads

Thread	Drill	Thread	Drill	Taper Pipe	Drill
#0-80	$3/64$	#12-24	#17	$1/8$	R
#1-64	#53	#12-28	#15	$1/4$	$7/16$
#1-72	#53	$1/4$-20	#8	$3/8$	$37/64$
#2-56	#51	$1/4$-28	#3	$1/2$	$23/32$
#2-64	#50	$5/16$-18	F	$3/4$	$59/64$
#3-48	$5/64$	$5/16$-24	I	1	$1 5/32$
#3-56	#46	$3/8$-16	$5/16$	$1 1/4$	$1 1/2$
#4-40	#43	$3/8$-24	Q	$1 1/2$	$1 47/64$
#4-48	#42	$7/16$-14	U	2	$2 7/32$
#5-40	#39	$7/16$-20	W		
#5-44	#37	$1/2$-13	$27/64$		
#6-32	#36	$1/2$-20	$29/64$		
#6-40	#33	$9/16$-12	$31/64$		
#8-32	#29	$9/16$-18	$33/64$		
#8-36	#29	$5/8$-11	$17/32$		
#10-24	#25	$5/8$-18	$37/64$		
#10-32	#21				

Rod and bolt sizes are O.D. (outside measurement).
Pipe sizes are I.D. (inside measurement).

Basin Wrench

The basin wrench shown in Fig. 2-34 is used primarily for loosening and tightening coupling nuts in hard-to reach points behind sinks

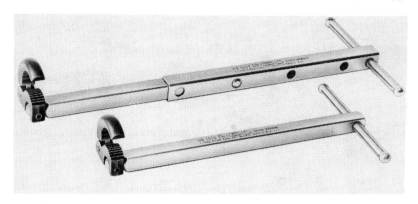

Fig. 2-34.   A basin wrench. *(Courtesy Ridge Tool Co.)*

and lavatories where it is impossible to use a standard wrench. The jaws are made to swivel 180° for tightening or loosening action; the jaws are spring loaded to provide a ratcheting action. Some brands have a telescoping shank for adjustment to best working length.

Hacksaw

Hacksaws are used by plumbers and pipe fitters for sawing many different materials. There are many occasions when piping must be cut in place and no other tool will serve the purpose. The hacksaw blade in this type saw can be mounted vertically or horizontally for cuts in difficult places. The saw shown in Fig. 2-35 has storage space for six extra blades in the backbone of the frame.

Fig. 2-35.   A hacksaw. *(Courtesy Ridge Tool Co.)*

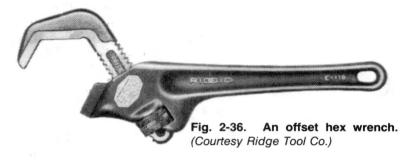

Fig. 2-36. An offset hex wrench.
(Courtesy Ridge Tool Co.)

Offset Hex Wrench

The extra-wide-opening smooth jaws of the wrench shown in Fig. 2-36 make it especially useful when working with chromium plated valves or fittings. The offset jaw gives easy access to nuts and fittings in hard-to-get-at places.

Fig. 2-37. A core-drilling machine.
(Courtesy Ridge Tool Co.)

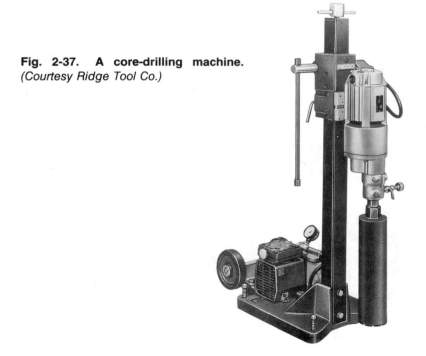

Miscellaneous Power Tools

The core-drilling machine shown in Fig. 2-37 is particularly useful on remodeling projects where holes must be drilled through masonry walls or concrete floors. The drills used with this machine have diamond-segmented bits which can drill through steel rods, etc., and bits are available for drilling holes from ½ to 14 in. diameter.

The copper-cleaning machine shown in Fig. 2-38 is used for fast cleaning and preparation of ½ in. through 4 in. copper tubing.

Drain-cleaning machines of the type shown in Fig. 2-39 are used to cut out roots, impacted debris or grease from drain or sewer piping. The machine shown has safety controls and is reversible.

Roll-grooving machines are used to roll grooves at the ends of piping to prepare the piping for mechanical couplings or other fit-

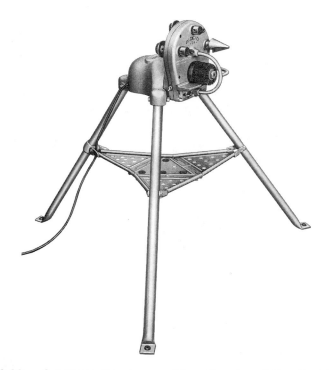

Fig. 2-38.   A copper-cleaning machine. *(Courtesy Ridge Tool Co.)*

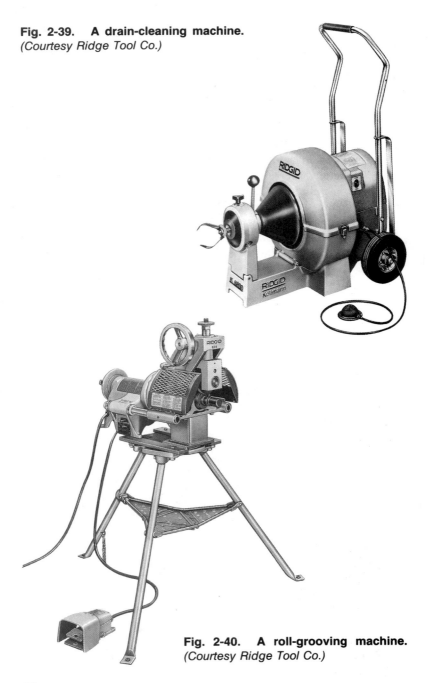

Fig. 2-39.   A drain-cleaning machine.
(Courtesy Ridge Tool Co.)

Fig. 2-40.   A roll-grooving machine.
(Courtesy Ridge Tool Co.)

tings. When mechanical fittings are used, threading is eliminated. One of the principal uses of grooved piping is in the installation of sprinkler systems for fire protection. A roll groover is shown in Fig. 2-40.

CHAPTER 3

Pipe Fittings and Valves

Since pipe cannot be obtained in unlimited lengths, and in practically all pipe installations there are numerous changes in directions, branches, etc., pipe fittings have been devised for the necessary connections. By definition, the term *pipe fitting* is used to denote all fittings that may be attached to pipes:

1. To alter the direction of a pipe.
2. To connect a branch with a main.
3. To close an end.
4. To connect two pipes of different sizes.

There are many different types of pipe fittings. These various fittings may be classed,

1. With respect to material, as:
   a. Cast iron.
   b. Malleable iron.
   c. Brass.
   d. Copper.
   e. Steel (cast and forged).
   f. Plastic (nonmetallic).
   g. Glass.
2. With respect to design, as:
   a. Plain.
   b. Beaded.
   c. Band.

3. With respect to the method of connecting, as:
   a. Screwed.
   b. Flanged.
   c. Hub-and-spigot.
   d. Cement.
   e. Soldered.
   f. Glued.
   g. Welded.
   h. Compression.
4. With respect to strength, as:
   a. Standard.
   b. Extra heavy.
   c. Double extra heavy.
5. With respect to the surface, as:
   a. Black, malleable or cast iron.
   b. Galvanized, malleable or cast iron.
6. With respect to finish, as:
   a. Rough.
   b. Semifinished.
   c. Polished.
7. With respect to service, as:
   a. Gas.
   b. Steam.
   c. Hydraulic (heavy pressure).
   d. Drainage.
   e. Railing.
   f. Sprinkler.
   g. Water.

The following definitions relating to pipes, joints, and fittings will be helpful to the pipe fitter and those desiring to acquire a knowledge of the subject.

Definitions

Bonnet—A cover used to guide and enclose the tail end of a valve spindle or a cap over the end of a pipe.

Branch—The outlet or inlet of a fitting not in line with the run, but which may make any angle.

Branch Ell—Used to designate an elbow having a back outlet in line with one of the outlets of the "run." It is also called a heel-outlet elbow, and is incorrectly used to designate a side-outlet or back-outlet elbow.

Branch Pipe—A very general term used to signify a pipe equipped with one or more branches. Such pipes are used so frequently that they have acquired common names such as tees, crosses, side- or back-outlet elbows, manifolds, double-branch elbows, etc. The term *branch pipe* is generally restricted to those pipes that do not conform to usual dimensions.

Branch Tee—A tee having many side branches. *See* Manifold.

Bull-head Tee—A tee, the branch of which is larger than the run.

Bushing—A pipe fitting for the purpose of connecting a pipe with a fitting of a larger size, being a hollow plug with internal and external threads to suit the different diameters.

Close Nipple—A nipple that is about twice the length of a standard pipe thread and is without any shoulder.

Coupling—A threaded sleeve used to connect two pipes. Commercial couplings are threaded to suit the exterior thread of the pipe. The term *coupling* is occasionally used to mean any jointing device and may be applied to either straight or reducing sizes.

Cross—A pipe fitting with four branches arranged in pairs, each pair on one axis, and the axes at right angles. When the outlets are arranged otherwise the fittings are branch pipes or specials.

Crossover—A fitting with a double offset, or shaped like the letter U with the ends turned out. It is made only in small sizes and is used to pass the flow of one pipe past another when the pipes are in the same plane.

Crossover Tee—A fitting made along the lines similar to a crossover, but having at one end two openings in a tee head, the plane of which is at right angles to the plane of the crossover bend.

Cross Valve—A valve fitted on a transverse pipe so as to open communication at will between two parallel lines of piping.

Crotch—A fitting that has a general shape of the letter Y. Caution should be exercised not to confuse the crotch and the wye (Y).

Drop Ell (or Wing Ell)—A 90° ell with extensions cast on the side for anchoring to wall surface.

Drop Tee—One having the same peculiar wings as the drop elbow.

Dry Joint—A joint made without gasket, packing, or compound of any kind, as a ground joint.

Elbow (ell)—A fitting that makes an angle between adjacent pipes. The angle is always 90° unless another angle is stated. *See* Branch, Service, and Union Ell.

Extra Heavy—When applied to pipe, means pipe thicker than standard pipe; when applied to valves and fittings, indicates units suitable for a working pressure of 250 lbs. per sq. in.

Header—A large pipe into which one set of boilers is connected by suitable nozzles or tees, or similar large pipes from which a number of smaller ones lead to consuming points. Headers are often used for other purposes—for heaters or in refrigeration work. Headers are essentially branch pipes with many outlets, which are usually parallel. Largely used for tubes or water-tube boiler.

Hub-and-Spigot Joint—The usual term for the joint in cast-iron pipe. Each piece is made with an enlarged diameter or hub at one end into which the plain or spigot end of another piece is inserted when laying. The joint is then made tight by cement, oakum, lead, rubber, or other suitable substance, which is driven in or caulked into the hub and around the spigot. Applied to fittings or valves, the term means that one end of the run is a "hub" and the other end is a "spigot," similar to those used on regular cast-iron pipe.

Hydrostatic Joint—Used in large water mains, and where sheet lead is forced tightly into the bell of a pipe by means of the hydrostatic pressure of a liquid.

Lead Joint—Generally used to signify the connection between pipes made by pouring molten lead into the annular space between a bell and spigot and then making the lead tight by caulking.

Lead Wool—A material used in place of molten lead for making pipe joints. It is lead fiber, about as coarse as fine excelsior, and when made in a strand, it can be caulked into the joints, making them very solid.

Lip Union—A special form of union characterized by a lip that prevents a gasket from being squeezed into the pipe to obstruct the flow. It is a ring union unless flange is specified.

Manifold—A fitting with numerous branches used to convey fluids between a large pipe and several smaller pipes. *See* Branch Tee. A header for a coil.

Medium Pressure—When applied to valves and fittings, means suitable for a working pressure of from 125 to 175 lbs. per sq. in.

Needle Valve—A valve provided with a long tapering point in place of the ordinary valve disk. The tapering point permits fine graduation of the opening. At times called a *needle-point* valve.

Nipple—A short piece of pipe, threaded on both ends, can be any length from a close nipple up to and including a 12-in. nipple. Pipe over 12 in. is regarded as cut pipe.

Reducer—A fitting with a female pipe thread on each end, and one thread is one or more pipe sizes smaller than the other; it is technically a reducing coupling. The larger size is always correctly named first, as: 1½″ × 1¼″ reducer. While other fittings, ells, tees, and wyes can be used to reduce pipe sizes, the term reducer is correctly applied only to a reducing coupling.

Roof Increaser—A fitting designed to increase the size of a waste or vent stack at the point where it exits the building at the roof line. *See* Fig. 6-1.

Run—A length of pipe made up of more than one piece of pipe. The portion of any fitting having its end "in line" or nearly so, in contrast to the branch or side opening, as of a tee.

Service Pipe—A pipe connecting mains with a building.

Shoulder Nipple—The next length to a close nipple. The actual amount of unthreaded pipe between the two threads will vary with the pipe size.

Standard Pressure—A term applied to valves and fittings suitable for a working steam pressure of 125 lbs. per sq. in.

Street Elbow—An elbow having one male thread and one female thread.

Street Tee—A tee having one male thread (end) and two female threads (one end and side).

Tee—A fitting that has one side outlet at right angles to the run. A bullhead tee is one in which the side outlet is larger than the run.

Union—A fitting used to connect pipes. There are two common types of unions: the flange union (Fig. 3-16) and the ground joint union (Fig. 3-7).

Union Ell—Union Ells are a combination union and elbow, in one fitting. They are made in both male and female patterns, and in 90° and 45° ells. (They are shown in Fig. 3-17.)

Union Tee—Union Tees are a combination union and tee in one fitting. They are made in both run and outlet patterns as shown in Fig. 3-17. Union tees are not a commonly used fitting.

Wye—A fitting that has a side outlet at 45°.

Cast-Iron Fittings

Standard beaded or flat-band fittings of cast iron are suitable for 125 lbs. of steam or 175 lbs. of water pressure. These fittings will actually require from 1000 to 2500 lbs. of pressure to burst them. The larger factor of safety is necessary in their use because of the strain due to expansion, contraction, weight of piping, settling, and water hammer. For steam pressures above 125 lbs., extra-heavy fittings should be used.

Malleable-Iron Fittings

Standard beaded or flat-band fittings of malleable iron are intended for working steam pressure up to 150 lbs. Such fittings have, at various times, been subjected to hydraulic pressures of from 2000 to 4000 lbs. without bursting. It would therefore seem possible that they would be safe for at least 250 lbs. of working steam pressure. If proper care is exercised in fitting and using them, they will undoubtedly be found satisfactory for working pressure up to 500 lbs. However, since all fittings are subjected to strain due to expansion, contraction, and making up the joints, they are not recommended for working pressures over 150 lbs. In fact, since extra-heavy fittings cost only a little more, it is not economical to use standard fittings for working pressures near 150 lbs. Standard plain-pattern malleable fittings are used for low-pressure gas and water and house plumbing.

Brass Fittings

Brass fittings are made in standard, extra-heavy, and cast-iron patterns (iron-pipe sizes), and are used for brass feed-water pipes where hard water makes steel pipes undesirable. The standard brass fittings are usually made in sizes from ¼ to 3 in., suitable for 125 lbs. working pressure; extra-heavy fittings, ⅛ to 6 in., suitable for 150 lbs. working pressure; cast-iron patterns in all sizes, suitable for 250 lbs. working pressure.

Cast-Steel Fittings

Cast-steel fittings are made extra heavy with screwed or flanged ends. The screwed fittings are listed in sizes from 3 to 6 in. The 3- to 4½-in. sizes (inclusive) are tested for 1500 lbs. hydrostatic pressure, and the 5- and 6-in. sizes for 1200 lbs. pressure. The radii of these fittings are larger than ordinary fittings, thereby reducing friction. They are suitable for the working pressures given when used in hydraulic installations in which shock is absent or so slight as to be negligible.

Ordinarily, these fittings, when subject to shock, should not be used for working pressures higher than 65 percent of the hydrostatic test pressure, and where shock is severe, 50 percent, or even 40 percent, will be more conservative. Installations of this character should always be protected by shock absorbers placed to the best advantage.

Forged-Steel Fittings

The extra-heavy hydraulic forged-steel screwed fittings are suitable for superheated steam up to 2350 lbs. working pressure, a total temperature of 899° F, and for cold water or oil working hydrostatic pressures up to 3000 lbs. They are regularly made from solid forgings in sizes ranging from ½ to 2½ in. inclusive, and are tested to 3000 lbs. hydraulic pressure. The double extra-heavy pattern is suitable for cold water or oil working hydrostatic pressures up to 6000 lbs. They are regularly made from solid forgings in sizes ranging from ⅜ to 2 in. inclusive, and are tested to 6000 lbs. hydrostatic pressure.

Various Fittings

There are a great many fittings due to the many modifications of each class of fittings and the several weights and different metals used. A list of these fittings may be divided into several groups, classified with respect to their use:

1. Extension or joining.
   a. Nipples.
   b. Lock nuts.

        c. Couplings.
        d. Offsets.
        e. Joints.
        f. Unions.
   2. Reducing or enlarging.
        a. Bushings.
        b. Reducers.
   3. Directional.
        a. Offsets.
        b. Elbows.
        c. Return bends.
   4. Branching.
        a. Side-outlet elbows.
        b. Back-outlet return bends.
        c. Tees.
        d. Y branches.
        e. Crosses.
   5. Shutoff or closing.
        a. Plugs.
        b. Caps.
        c. Flanges.
   6. Union.
        a. Union elbows.
        b. Union tees.

Nipples

By definition, a nipple is a piece of pipe up to and including 12 in. in length, threaded on both ends; pipe over 12 in. long is regarded as cut pipe. With respect to length, nipples may be classed as:

   1. Close.
   2. Short.
   3. Long.

Where fittings or valves are to be very close to each other, the intervening nipple is just long enough to take the threads at each end, being called a *close nipple*. If a small amount of pipe exists between the threads, it is called a *shoulder* or *short nipple*, and where a larger amount of bare pipe exists, it is called a *long nipple* or *extra-long nipple*.

Fig. 3-1.   A right- and left-hand threaded nipple.

Nipples having a right-hand thread on one end and a left-hand thread on the other are generally used in steam heating piping instead of unions. Fig. 3-1 shows such a nipple, with a hexagon nut at the center forming part of the nipple.

Lock Nuts

Lock nuts are made with faced end for use on long screw nipples having couplings, and with a recessed or grooved end to hold packing where this is depended on to make a tight joint. The use of lock nipples should be avoided wherever possible since the joint is not as good as that obtained by a union.

Couplings

An ordinary coupling, shown in Fig. 3-2, is furnished with pipe, one coupling per length, with R.H. threads. An extension coupling is made with a male thread on one end and a female thread on the other. It is used primarily for extending an existing opening to a new wall line; for instance, if ceramic tile is added to an existing wall, an extension coupling added to the existing fitting will extend the existing piping to the same relative distance behind the new wall line as it was originally. An extension coupling is shown in Fig. 3-3.

Fig. 3-2.   A pipe coupling.

Fig. 3-3.   An extension coupling.

Fig. 3-4.   A pipe reducer.

A reducing coupling or reducer, Fig. 3-4, is made in two styles: concentric, with the centers on the same plane, and eccentric, with the top (or bottom) on the same plane.

Unions

There are various kinds of unions. The plain union, as shown in Fig. 3-5, requires a gasket; the two pipes to be joined by the union must be in alignment to secure a tight joint because of the flat surfaces that must press against the gasket. This limitation is shown in Fig. 3-6.

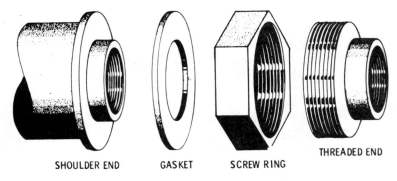

SHOULDER END   GASKET   SCREW RING   THREADED END

Fig. 3-5.   A gasket-type union.

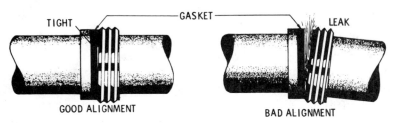

Fig. 3-6.   The importance of aligning unions.

GROUND-JOINT
UNION

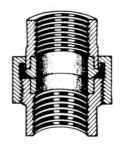

Fig. 3-7.   A ground-joint union.

To avoid this difficulty, and also to avoid the inconvenience of the gasket, various unions having spherical seats and ground joints have been devised. These consist of a composition ring bearing against iron, or with both contact surfaces of composition. Fig. 3-7 shows the construction of a *ground-joint union*. The joint has spherical contact, and the illustration shows the tight joint secured. Unions are also made entirely of brass with ground joints.

Right and Left Nipples—Right and left nipples are made to serve as unions, primarily in steam coils. Right and left nipples must be screwed into a right and left fitting. Right and left nipples have largely been replaced by newer methods of connection.

Bushings

Bushings are often confused with reducers. The function of a bushing is to *connect the male end of a pipe to a fitting of larger size*. It consists of a hollow plug with male and female threads to suit the different diameters. A bushing may be regarded as a reducing fitting. As generally manufactured, bushings 2½ in. and smaller, reducing one size, are malleable iron; reducing two or more sizes are cast iron; all above 2½ in. are cast iron, except brass bushings, which may be obtained in sizes from ¼ to 4 in.

Bushings are listed by the *pipe size of the male thread*; thus a ¼-in. bushing joins a ¼-in. fitting to a ⅛-in. pipe. To avoid mistakes, however, it is better to specify the size of both threads, for instance, calling the bushing just mentioned a ¼-in. × ⅛-in. bushing. The regular-pattern bushing has a hexagon nut at the female end for screwing the bushing into the fitting. The *faced* bushing is used for very close work, having a faced end in place of the hexagon nut.

Fig. 3-8.   Various types of bushings.

This may be used with a long screw pipe and faced lock nut to form a tight joint or to receive a male end fitting for close work. Fig. 3-8 shows the plain and faced types of bushing. As shown in Fig. 3-8, bushings are made in several different patterns: concentric hex and face bushings and eccentric hex bushings.

Concentric bushings are used when it is necessary to keep both sizes of piping on the same center.

Eccentric bushings are used when it is desirable to keep the top plane of both sizes of piping as near level as possible. At times, for instance in a hot-water heating system, it is important that both sizes of piping be on the same plane in order to eliminate air problems.

Return Bends

Return bends are largely used for making up pipe coils for steam heating and for water boilers. They are U-shaped fittings with a female thread at both ends and are regularly made in three patterns, known as:

1. Close.
2. Medium.
3. Open.

Some manufacturers also make an extra-close and an extra-wide pattern. These patterns represent various widths between the two arms. There seems to be no standard as to the spacing of the arms for the different patterns; hence, for close work, the fitter should ascertain the center-to-center dimensions from the manufacturer's catalogue of the make to be used. Table 3-1 gives the dimensions of the various elbows. It should be noted that return bends are made in ¼ in. and ⅜ in. sizes.

For making up so-called coils from short lengths of pipe, return bends may be obtained with elbows, as illustrated in Fig. 3-9. The pipes, when screwed into the fitting, will not be parallel but will

Table 3-1 Malleable Return Bends

Size	Extra Close Center to Center	Weight per 100 (plain)	Close Center to Center	Weight per 100 (banded)	Medium Center to Center	Weight per 100 (banded)	Open Center to Center	Weight per 100	Wide Center to Center
1/4	....	....	3/4	15.5	....	....	1 1/8	21	....
3/8	....	....	7/8	22	....	....	1 1/4	22	....
1/2	....	....	1 1/8	35	1 1/4	34	1 1/2	44	....
3/4	1 1/4	77	1 3/8	71	1 9/16	60	2	83	6
1	1 9/16	92	1 3/4	100	1 7/8	92	2 1/2	140	3, 4, 4 1/2, 5, 6, 7, 8
1 1/4	....	....	2 1/8	168	2 1/4	160	3	200	4, 5, 6, 9
1 1/2	....	....	2 1/2	244	2 9/16	255	3 1/2	310	5, 6, 8
2	....	....	2 3/4	388	3 1/16	337	4 3/8	550	2, 6, 7, 8
2 1/2	....	....	3 7/8	631	....	....	4 3/4	710	....
3	....	....	4 1/2	880	....	....	6 1/4	1050	....
3 1/2	....	....	5	1400	....	....	6 1/2	1550	....
4	....	....	....	....	....	....	7	1850	....
5	....	....	6		....	....	....	....	....

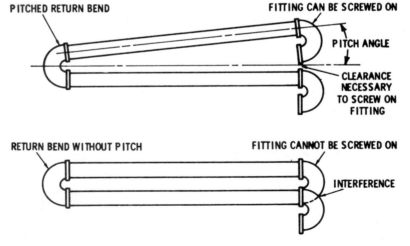

Fig. 3-9. Coils made with short pipe lengths and various angle elbows.

spread like the sides of the letter V. Such bends are usually listed for which the pitch is suitable.

Side-Outlet Elbows

The two openings of an elbow indicate its *run*, and when there is a third opening, the axis of which is at 90° to the plane of the run, the fitting is a side-outlet elbow, as shown in Fig. 3-10. These fittings are regularly made in sizes ranging from ¼ in. to 2 in. inclusive, with all outlets of equal size and with the side outlet one and two sizes smaller than the rim outlets. In general, it is not wise to specify fittings of this kind, which are not in much demand, unless the more usual forms are difficult to find.

Back- and Side-Outlet Return Bends

These types of bends are simply return bends provided with an additional outlet at the back or side, as shown in Fig. 3-11. They are regularly made in sizes ranging from ¾ in. to 3 in. inclusive, in the close or open patterns.

Fig. 3-10.   Cast-iron elbows with side outlets.

Fig. 3-11.   Cast-iron return bends with back and side outlets.

Tees

Tees are the most important and widely used of the branching fittings. Tees, like elbows, are made in many different sizes and patterns. They are used for making a branch of 90° to the main pipe, and always have the branch at right angles. When the three outlets are of the same size, the fitting is specified by the size of the pipe, as a ½-in. tee; when the branch is a different size than the run outlets, the size of the run is given first, as a 1″ × ¼″ tee. When all three outlets are of different sizes, they are all specified, giving the sizes of the run first as a 1½″ × 1¼ × 1″ tee.

Wyes

A wye is a fitting with the side opening set at an angle of 45°. Wyes are made in both single and double patterns, and are made in straight and reducing sizes. The correct way to "read" a wye is: end, end, and side. Thus a wye with all openings being 1½ in. is a 1½-in. wye; a wye with 1½-in. end openings and a 1¼-in. side opening is a 1½″ × 1½″ × 1¼″ wye.

Crosses

A cross is simply an ordinary tee having a back outlet opposite the branch outlet (Fig. 3-12). The axes of the four outlets are in the same plane and at right angles to each other. Crosses, like tees, are made in a number of sizes. Regarding a cross as a tee with a back outlet, the tee part is made in various combinations of sizes, similar to ordinary tees, but the back outlet is always the same size as the opposite side outlet of the tee parts.

Plugs

A plug is used for closing the end of a fitting having a female thread. Metal plugs are made of steel, cast iron, and brass, in patterns as

Fig. 3-12.   Two cross connectors.

Fig. 3-13.   Various pipe plugs.

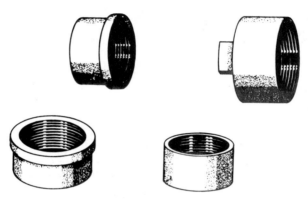

Fig. 3-14.   Various pipe caps.

shown in Fig. 3-13, with either raised square heads or recessed square heads. Plugs are also made of plastics, PVC, and ABS.

Caps

A cap is used for closing the end of a pipe or fitting having a male thread. Caps, like plugs, are made of cast iron, malleable iron, and brass. Fig. 3-14 shows various cap designs. Plain and flat-band or beaded caps are regularly made in sizes from ⅛ in. to 6 in. inclusive; cast-iron caps from ⅜ in. to 15 in. inclusive.

Flanges

Flanges are made of copper and brass for use with copper piping and of steel for use with steel or cast-iron piping and equipment (Fig. 3-15). Flanges are made for use with threaded piping, for welded piping (slip-on and weld neck types), and solder type, for

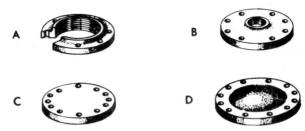

Fig. 3-15. Various flanges: (A) threaded flange; (B) threaded reducing flange; (C) and (D) blind flanges.

soft or hard soldering. Flanges are made in straight pipe sizes for use when bolting piping and equipment together, and are also made in reducing sizes. They are also made for closing the end of a run of piping; this type is called a *blind flange*.

Flanged Fittings and Valves——Flanged fittings and valves are commonly used on piping 2½ in. and larger (Fig. 3-16). The use of flanged fittings and valves permits easy removal of pumps, specialized equipment, etc. Flanged fittings and valves are used on water, gas, steam, hot water, air, air conditioning, and sewage piping, as well as other specialized piping. Flanges used with these fittings are called *companion* flanges because they are made in standard sizes to mate with standard valves and fittings.

Union Elbow and Tees

The frequent use of unions in pipe lines is desirable for convenience in case of repairs. Where the union is combined with a fitting, the advantage of a union is obtained with only one threaded joint instead of two, as in the case of a separate union. A disadvantage of union fittings is that they are not usually as easily obtainable as ordinary fittings. Fig. 3-17 shows various union elbows and union tees of the female and of the male and female types.

Drainage Fittings

Drainage fittings of cast iron, copper, and plastic differ slightly from ordinary fittings. They are so constructed that when the pipe is made into the fitting, there is a smooth, unobstructed passageway

TYPES OF FLANGED FITTINGS

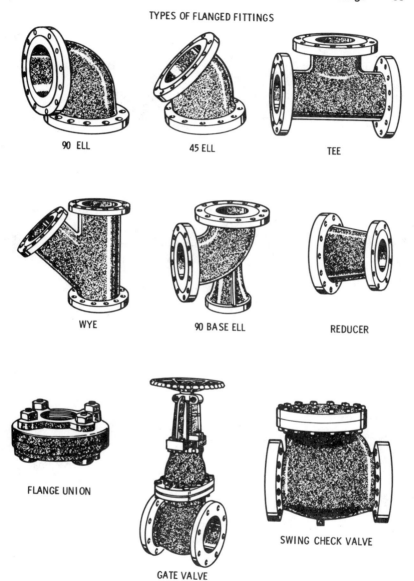

90 ELL

45 ELL

TEE

WYE

90 BASE ELL

REDUCER

FLANGE UNION

GATE VALVE

SWING CHECK VALVE

Fig. 3-16. Flanged fittings and valves.

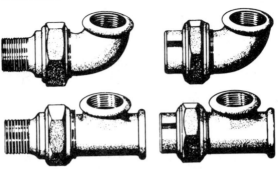

Fig. 3-17.   Various union elbows and union tees.

for liquid and water-carried wastes. If standard water fittings were used for drainage, there would be a shoulder at the point where the pipe stops in the fitting, and this shoulder would obstruct the flow through the pipe. Copper and plastic fittings are better for drainage purposes because they extend completely into the fitting recess; when one uses threaded fittings, the pipe may not screw into the recess completely when tight and may leave a shoulder or ridge. Drainage fittings are shown and compared with a standard fitting in Fig. 6-9.

Valves

Valves are used in a piping system to control the flow of liquids, gas, or air. There are many different types of valves; some of the commonly used types and the reasons for their use are explained in this chapter.

Faucets are also valves, but they are valves located at the end of a main or branch and serve to control the flow of liquids at a terminal point or at a fixture.

Gate valves are most commonly used in industrial and commercial applications as stop valves, to turn on or to stop flow as opposed to regulating flow. They are usually used to control a main, branch, or an item of equipment. A gate valve has a full-size passageway when fully open and is normally used in either a full open or full closed position. Gate valves are best suited for full open flow because the fluid moves through them in a straight line, nearly without resistance or pressure drop, when the disc is raised from the waterway. Gate valves are not intended for throttling the flow or for frequent operation. If the valve is only partially open, it may vibrate or chatter and cause damage to the seating surface. Repeated movement of the disc near the point of closure will cause the seating surfaces to rub together until they are galled or scored on the downstream side. There are two basic types of gate valves: *rising stem* and *nonrising stem.*

Rising stem valves are generally installed where space is available and in conditions where the visual determination of whether the valve is open or closed is important.

Nonrising stem valves are generally installed where space is limited. Because the disc rides up and down on the stem and the stem rotates in the packing, the life of the packing is greatly increased. However, it is not possible to determine visually whether a nonrising stem valve is open or closed.

Gate valves are made with solid wedge discs (Fig. 3-18) and double wedge discs (Fig. 3-19). The flanged valve (Fig. 3-18) is an O.S. & Y. (open screw and yoke) type.

Globe Valves

The most familiar type of valve is the globe valve, which is extensively used in piping systems for water, air, and steam. This type of valve is designed to be placed in the run of a pipe line. As shown in the cutaway view in Fig. 3-20, a globe valve has a spherical casting with an interior partition that shuts off the inlet from the outlet except through a circular opening in the seat. Screwed into an opening in the top of the casting is a plug having a stuffing box and a threaded sleeve in which the valve spindle operates. On the lower end of this spindle is the valve proper, and on the other end is a handwheel. The valve is closed by turning the handwheel clockwise, which lowers the spindle and valve until it presses firmly and evenly on the valve seat, thus closing communication between the inlet and outlet. By turning the handwheel in the opposite direction (counterclockwise), the valve is opened.

Globe valves are primarily control valves because of their throttling characteristics. The disc travel is shorter, and they generally operate with fewer turns of the handwheel than that of a gate valve. The maintenance of these valves is most economical since they do

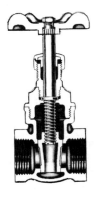

**GATE VALVE
NON-RISING STEM—
SOLID WEDGE DISC**

(A) I.P.S. (threaded).

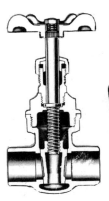

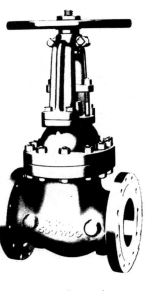

(C) Flanged.

**GATE VALVE
NON-RISING STEM—
SOLID WEDGE DISC**

(B) Copper to copper (sweat).

Fig. 3-18.   Gate valves.

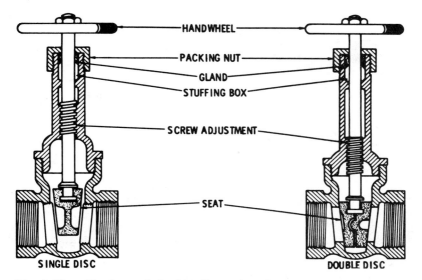

Fig. 3-19.   Single- and double-disc gate valves.

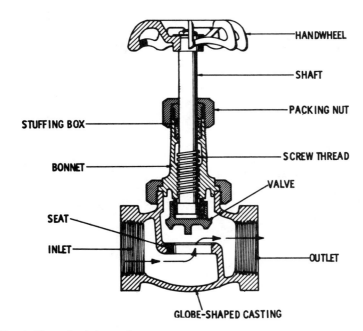

Fig. 3-20.   A globe valve.

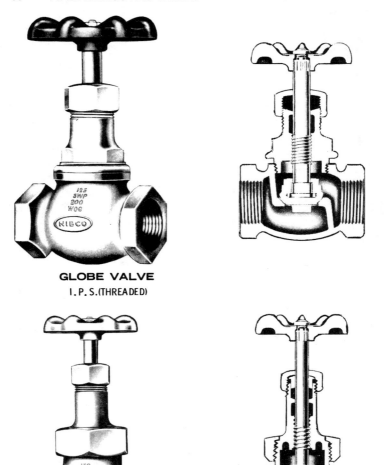

GLOBE VALVE
I. P. S.(THREADED)

GLOBE VALVE
COPPER TO COPPER (SWEAT)

Fig. 3-21.   Two types of globe valves.

FLAT ← VALVE SEAT → BEVELED

Fig. 3-22. A flat and a beveled valve and valve seat.

not need to be removed from the pipe line for disc replacement. Steam discs and W.O.G. (water-oil-gas) discs are securely fastened to a swivel-type holder to insure even wear and uniform seating. Globe valves should be installed with the pressure on top of the disc, whenever conditions permit, so that the line pressure can add to the seating pressure. This is particularly useful in hot-water and steam applications where subsequent cooling of the valve may cause leakage to occur due to the different cooling rates of the various parts of the valve. Service conditions often make installation with the pressure under the seat desirable; and in some cases, such as boiler-fed water lines, such installation is mandatory by codes or regulations.

Fig. 3-21 shows two types of globe valves: the I.P.S. (threaded) and the copper to copper (sweat). The seat and valve may have their contact surfaces either flat or beveled, as shown in Fig. 3-22. The valve disc may be of metal or fiber. Fiber seats should be interchangeable. A globe valve will remain leakproof longer than a gate valve. An objection is that unless properly designed, the opening through the seat of the valve is not the full area of the pipe size;

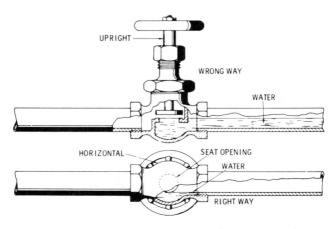

Fig. 3-23. The proper way to install a globe valve.

this and the contorted passages offer considerable resistance to water flow. A serious objection to the use of globe valves on water lines that must be drained in freezing weather is that it is impossible to drain the water from a horizontal line when the valve stem is in an upright position. A globe valve should be installed with the stem in a horizontal position as shown in Fig. 3-23.

Angle Valves

An angle valve is a valve with the inlet and outlet at 90° to each other, as shown in Fig. 3-24. Such valves can serve the double purpose of controlling the flow and changing the direction of the pipe line, thus eliminating the need for an elbow. In connecting a globe valve, it is important to place it in the line so that its inlet side will carry the pressure when the valve is closed. Otherwise it will be impossible to repack the stuffing box while the line is under pressure.

Check Valves

There are many different check valves designed for different applications. Basically, all check valves are designed with the same purpose: to permit unrestricted forward flow and at the same time prevent backflow. The basic types of check valves are the swing

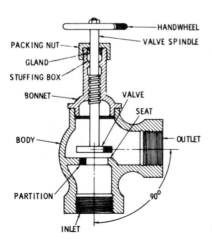

Fig. 3-24. An angle valve.

Y-CHECK VALVE
I.P.S. (THREADED)

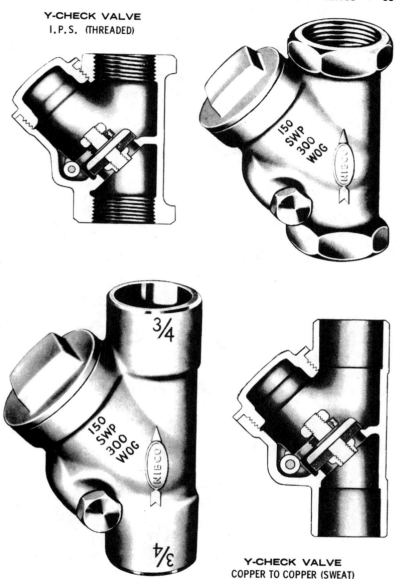

Y-CHECK VALVE
COPPER TO COPPER (SWEAT)

Fig. 3-25. Two common types of swing check valves.

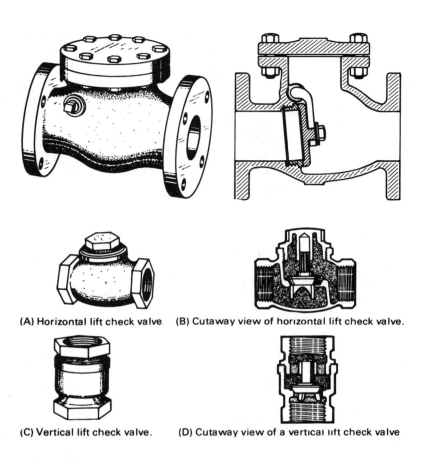

(A) Horizontal lift check valve. (B) Cutaway view of horizontal lift check valve.

(C) Vertical lift check valve. (D) Cutaway view of a vertical lift check valve.

CAP
BODY
DISC HOLDER
SNAP
RING
HINGE
PIN
HINGE DISC DISC NUT

Fig. 3-26. Exterior and cut-away views of check valves.

check, the horizontal lift check, and the vertical lift check. Common types of swing check valves are shown in Figs. 3-25 and 3-26. Generally speaking, swing check valves offer less resistance to flow than other types of check valves. Both horizontal and vertical lift check valves are shown in Fig. 3-26.

A swing check valve should be sufficiently large in diameter so that it will deliver the required amount of water without lifting the disc more than ⅛ in. Higher lifts result in rapid destruction of the valve seat from the hammering action of the valve, especially when used with engine-driven pumps. When the water feed is continuous, as when an injector is used, the valve remains off its seat and a higher lift is not objectionable.

Cross Valves

The essentials of a cross globe valve are shown in Fig. 3-27. This type of valve is used where it is desired to control the flow at the junction of a main line and a branch. In cases where the branch line is the inlet, the valve can be repacked while under pressure, but unfortunately the branch must frequently be made the outlet, in which case the valve cannot be repacked while under pressure.

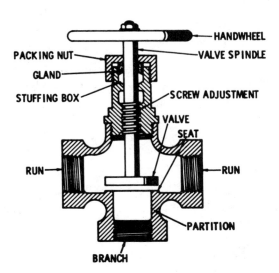

Fig. 3-27.   A cross globe valve.

Compression Stops

Compression stops are essentially globe valves and are commonly used on the supply piping to individual fixtures. Most plumbing codes require valves on the supply piping to individual fixtures so that when the faucet needs repair it will not be necessary to shut off the water to the entire building in order to make the repair. Compression stops are made in straight angle patterns, and for I.P.S. and sweat connections. They are also made in rough brass, polished brass, and chrome-plated finishes, depending on the usage of the stop. Compression stops are shown in Figs. 3-28 and 3-29.

Stop and Waste Valves

Stop and waste valves are used primarily to protect piping from freezing. A stop and waste valve is essentially a compression stop with a drain feature built into it. When the valve is closed and the button and rubber washer are removed, by turning the button counterclockwise, any water in the piping from the valve to the end of the piping will drain out, if the piping is open on the end. The common usage for a stop and waste valve is to protect a sillcock or hose bibb. It is necessary to open the sillcock or hose bibb to relieve any vacuum that may be caused by the partial drainage of water from the piping. Stop and waste valves are also made in a pattern called an automatic stop and waste. Automatic stop and waste valves have a port that opens when the valve is in closed position, thus the valve drains automatically. A hose bibb, sillcock, etc., which is on the end of the piping run, should also be opened to relieve a possible vacuum.

Needle Valves

A needle valve is a form of globe valve used where only a very small amount of flow and close regulation are desired. In place of a disc, the pointed end of the spindle forms the valve, which sets on a beveled seat of the same taper. The standard angle of the seat is 30° to the spindle axis. Fig. 3-30 shows the construction of a needle valve.

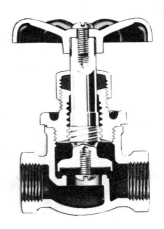

STOP VALVE
I. P. S. (THREADED)

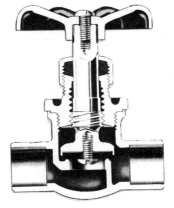

STOP VALVE
COPPER TO COPPER (SWEAT)

Fig. 3-28. Two types of straight compression stops.

 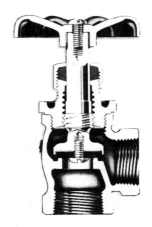

ANGLE STOP VALVE

 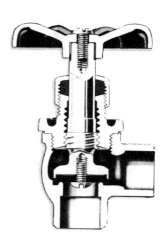

ANGLE STOP VALVE
COPPER TO COPPER (SWEAT)

Fig. 3-29.   Two types of angle compression stops.

Fig. 3-30.   A needle valve.

Pressure-Regulating Valves

Pressure-regulating valves (Fig. 3-31) are used for many purposes by the plumbing and pipe fitting trades. The principal uses are on water, air, and oil service piping. The basic principles of valve operation are the same regardless of the type of usage.

A pressure-reducing valve is designed to maintain a lower-than-main pressure to a building or to a fixture. If, for instance, the supply main pressure would be 125 psi it would be desirable to lower the building pressure to 60 psi to protect the piping and appliances. It is recommended by many manufacturers of commercial type dishwashing equipment that the working pressure of the dishwasher not be over 25 psi. A pressure-reducing valve can be

Fig. 3-31.   A pressure-reducing valve. *(Courtesy Watts Regulator Co.)*

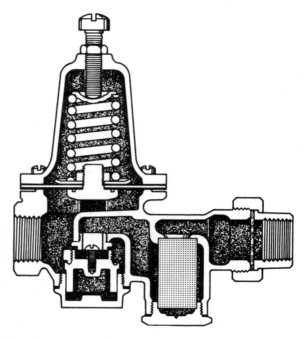

Fig. 3-32.   A cut-away view of a pressure-reducing valve with integral strainer.

installed between the booster heater and the dishwasher for this purpose. A pressure-reducing valve should have an integral strainer, Fig. 3-32, or a strainer should be installed on the high-pressure or inlet side of the valve.

Float Valves

The duty of a float valve is to shut off the water supply to a tank or receptacle when the water has reached a predetermined level. The automatic action is due to the rising level of the water during the filling of the tank, carrying up with it a float which, by suitable gearing, closes the water supply valve. When the water is discharged from the tank in flushing, the float descends by gravity, and the valve opens by pressure of the water supply.

The construction and operation of a typical float valve are shown in Fig. 3-33. An adjusting screw to establish the water level at which

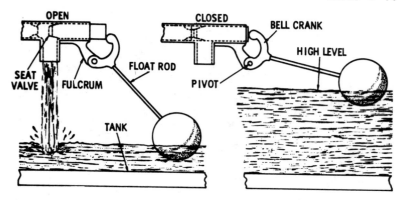

Fig. 3-33.   Float valve in open and closed positions.

the valve closes is provided. The principle of such adjustment is shown in Fig. 3-34, and use should be made of this means of adjustment when necessary rather than resorting to the objectionable practice of bending the float rod. Fig. 3-35 shows how the rod is bent to lower the level from *A* to *B*, and the possible result due to such practice.

A form of valve employing a diaphragm is shown in the open and closed position in Fig. 3-36. It acts by hydraulic pressure, depending on the differential area principle to keep it closed. In construction, a diaphragm valve divides the valve chamber into two compartments. In the lower compartment is the valve seat of the

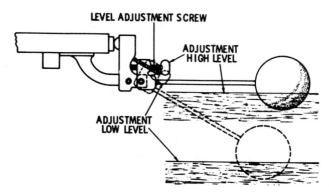

Fig. 3-34.   A float valve level adjustment.

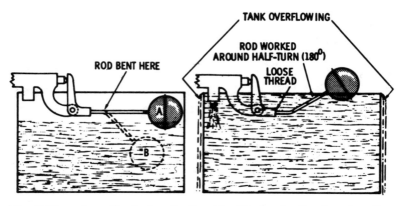

Fig. 3-35. A method of adjusting the float valve by bending the rod.

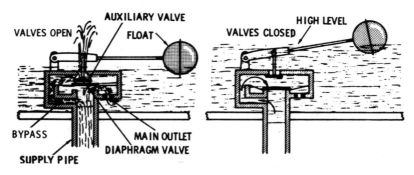

Fig. 3-36. Adjustment of a hydraulic float valve.

diaphragm valve, outlet ports, and a bypass passage leading to the upper compartment that has no opening except that of an auxiliary valve operated by the float as shown. This arrangement forms an efficient device and can be made to control the flow from a large tank with a small float. Fig. 3-37 shows the shank of the float valve and the method of securing a tight joint where the shank passes through the bottom of the tank.

This type of valve is specifically designed for use with hot water storage tanks and heaters.

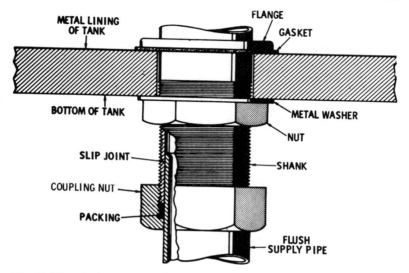

Fig. 3-37. The base of a fill valve.

Safety Relief Valves

Relief valves are safety devices designed to protect against pressure and/or temperature if for any reason run-away firing conditions should occur. Building codes require the installation of relief valves on domestic water heaters, on direct and indirect fired water storage tanks, on hot-water heating boilers, and on steam boilers. The type and pressure/temperature rating of the correct valve to use depends on the individual application. A relief valve used for any of these purposes should have a test lever, be self-closing, and bear a tag or label showing the ASME (American Society of Mechanical Engineers) code symbol and letters and the AGA (American Gas Assn.) symbol of approval. A typical temperature/pressure relief valve used on domestic water heaters is shown in Fig. 3-38. The valve shown in Fig. 3-39 is one type installed on hot-water and steam boilers.

Cocks

A cock is a valve intended to form a convenient means of shutting off the flow of water in a line. It is similar in construction to a

Fig. 3-38. A temperature and pressure relief valve. *(Courtesy Watts Regulator Co.)*

Fig. 3-39. An A.S.M.E. rated pressure relief valve. *(Courtesy Watts Regulator Co.)*

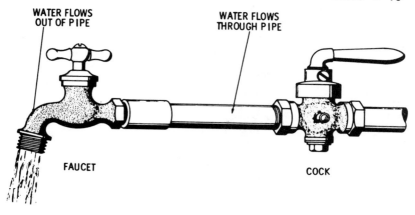

WATER FLOWS
OUT OF PIPE

WATER FLOWS
THROUGH PIPE

FAUCET

COCK

Fig. 3-40.   Difference between a faucet and a cock.

ground-key faucet but differs in that it is arranged to be placed *in the pipe line* instead of *at an outlet*. The distinction is shown in Fig. 3-40. To meet the various requirements of service, there are several kinds of cocks as follows:

1. Straight-way.
2. Three-way.
   a. Two-port.
   b. Three-port.
3. Four-way.
   a. Two-port.
   b. Three-port.
   c. Four-port.
4. Swing.
5. Waste or drain.
6. Corporation.

Straight-Way Cocks

Fig. 3-41 shows the construction of a straight-way cock, being virtually the same as a ground-key faucet except for the inlet ends and the detachable handle. The general appearance of several straight-way cocks (for steam) is shown in Fig. 3-42. It will be seen that there is a great variety of patterns to meet all requirements.

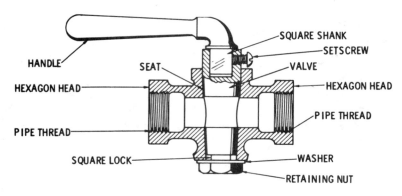

Fig. 3-41. A straight-way cock.

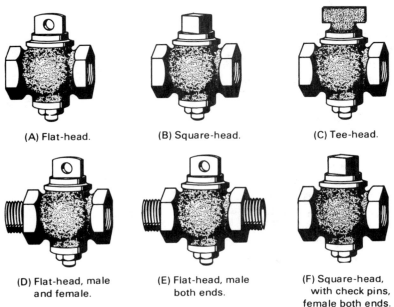

(A) Flat-head.

(B) Square-head.

(C) Tee-head.

(D) Flat-head, male and female.

(E) Flat-head, male both ends.

(F) Square-head, with check pins, female both ends.

Fig. 3-42. Various types of straight-way cocks.

It should be distinctly understood that the primary duty of a cock is to *control* rather than *regulate* the flow of water, that is, to shut off water from a pipe line in case of repairs or for draining in cold weather. In order to insure that the cock handle will be turned to the fully open or fully closed position, some units are provided with stops and a check pin. The stops are single projections on the body adjacent to the valve, and the check pin is inserted in the valve so that it will strike against the stops, limiting the angular movement.

Three-Way Cocks

Three-way cocks are used to control the flow at the junction of:

1. A main line and two branch lines.
2. A main line and one branch line.

For a main line and two branch lines, a two-port three-way cock is used, as shown in Fig. 3-43. The water may be directed to either branch or shut off from both. Where there is only one branch, the three-port cock permits control of the flow to the branch and to the run of the main line beyond the cock, as in Fig. 3-44.

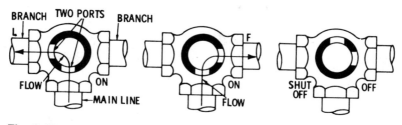

Fig. 3-43.   A two-port three-way cock.

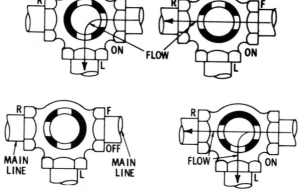

Fig. 3-44.   A three-port three-way cock.

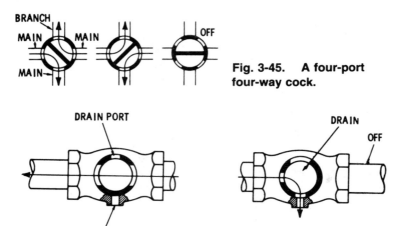

Fig. 3-45.   A four-port four-way cock.

Fig. 3-46.   A waste or drain cock, showing operations.

Four-Way Cocks

The range of flow control with a four-way cock is quite varied, as this pattern may be had with either two-, three-, or four-port valve, as shown in Fig. 3-45.

Waste or Drain Cocks

A waste or drain cock is used to drain a line from which the water is shut off. This is accomplished with a straight-way cock in which a small or auxiliary port is provided at right angles to the two main ports, and which is connected to a drain outlet in the side of the unit. The operation of this type cock is shown in Fig. 3-46.

Waste or drain cocks should always be used to protect exposed lines in freezing weather, so that they may be conveniently shut off and drained in one operation. Corporation cocks are special forms used to shut off the water supply from a city main to a house main.

Thermoplastic Valves

Thermoplastic valves are designed to be used in conjunction with plastic pipe, primarily for the handling of liquids, gases, or other materials which would corrode or damage metallic valves.

Those persons associated with the piping trades should be familiar with the thermoplastic valves shown in this chapter; additional information on these and other valves will be found in Chapter 2, Process Piping, Volume III.

Ball Valves

A ball valve performs an ON/OFF function and derives its name from the flow-controlling ball located within the body of the valve. The hole through the center of the ball provides a no-resistance path when the valve is in an open position. The valve requires only a one-quarter (90°) turn from full ON to full OFF. True Union valves are so named because the collars or nuts on each end of the valve can be loosened and the valve section removed for inspection or part replacement. The valve shown in Fig. 3-47 is precision machined in critical areas to maintain correct tolerances and prevent

Cut-away view

Fig. 3-47.   A True Union ball valve. *(Courtesy Hayward Industrial Products, Inc.)*

leakage. The valve is made of PVC/CPVC/Polypropylene with threaded or socket ends.

Ball Check Valves

Made of PVC/CPVC/Polypropylene with threaded, socket, or flanged ends, the check valve shown in Fig. 3-48 unseats to permit flow in one direction, but seats against a seal to prevent backflow. The True Union feature, the two collars or nuts on the ends of the valve, can be loosened to permit removal and inspection of the ball and seal surfaces.

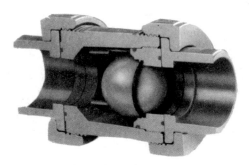

Cut-away view

Fig. 3-48. **A True Union check valve.** *(Courtesy Hayward Industrial Products, Inc.)*

Flanged Gate Valves

The flanged gate valves shown in Fig. 3-49 are designed for ON/OFF service and are also rated for full vacuum service. They are manufactured in sizes ranging from 1½ in. to 14 in. and in rising stem and non-rising stem types with PVC body and plug.

Rising stem type

Non-rising stem type

Fig. 3-49.   Two types of thermoplastic flanged gate valves. *(Courtesy Asahi/America, Inc.)*

Flanged globe valve

Socket-type globe valve   Threaded globe valve

Fig. 3-50.   Three types of thermoplastic globe valves. *(Courtesy Asahi/America, Inc.)*

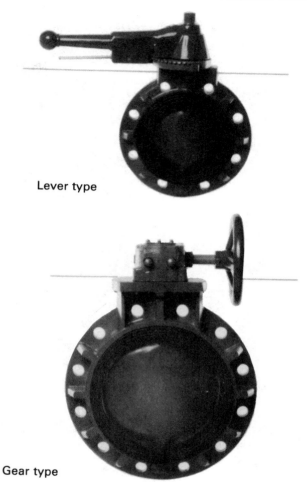

Lever type

Gear type

Fig. 3-51. Two types of thermoplastic flanged butterfly valves.
(Courtesy Asahi/America, Inc.)

Globe Valves

The flow through a globe valve follows a course that takes practically a 90° change of direction twice. But, because the seating of a globe valve is parallel to the line of flow of the liquid, it can be used frequently to throttle the flow of the liquid to any required degree or to give positive shutoffs. The turbulence caused by the change of direction through the valve does cause some line resistance and pressure drop, but the economy and dependability of this type valve make it desirable for many applications where this pressure drop is not critical.

Socket, threaded, and flanged type globe valves are shown in Fig. 3-50.

Butterfly Valves

Butterfly valves provide very accurate rate-of-flow control of materials passing through this type of valve. A lever-operated and a geared butterfly valve are shown in Fig. 3-51. Lever-operated butterfly valves range in size from 2 in. to 6 in.; geared valves range from 6 in. to 24 in.

CHAPTER 4

Cutting, Threading, and Installing Steel Pipe

A large part of plumbers' and pipe fitters' work consists of taking measurements, then cutting, threading, and installing black or galvanized steel pipe. When pipe is manufactured, all steel pipe (except stainless steel) is classed as black pipe. Galvanized pipe is made by first thoroughly cleaning black pipe, then coating it with zinc.

Pipe up to and including 4 in. I.D. (inside diameter) is usually cut and threaded on the job. Pipe larger than 4 in. I.D. is usually cut and threaded in the pipe shop. Steel pipe is always listed by I.D. size.

Pipe Cutting

Steel pipe received from the manufacturer comes in lengths varying from 20 to 22 ft. (usually 21 ft.), which makes it necessary to cut it to the proper length. This may be done with a hacksaw or a pipe cutter, the pipe in either case being held in a pipe vise such as the one shown in Fig. 2-8. Care should be taken in securing the pipe in the vise (especially when threading) that the jaws hold the pipe sufficiently firm to prevent slipping, but the clamp screw should not be tightened so tightly that the jaw teeth will dig excessively into the pipe.

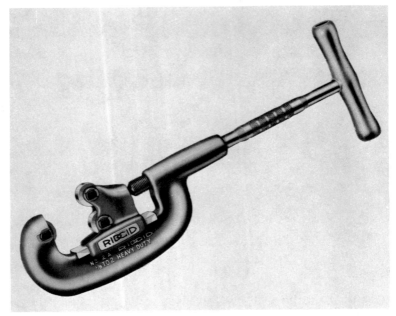

Fig. 4-1.   A single-wheel pipe cutter. *(Courtesy Ridge Tool Co.)*

Most pipe cutters consist of a hook-shaped frame on whose stem a slide is moved by a screw. One or more cutting discs or wheels are mounted on the slide and/or frame, and forced into the metal as the tool is rotated around the pipe. Pipe cutters may be classified as:

1. Wheel.
2. Combined wheel and roller.

A single-wheel pipe cutter is shown in Fig. 4-1. This model can be converted to a three-wheel cutter by replacing the two rollers with cutter wheels.

The single-wheel cutter is best for all-around work, but there are times when a three-wheel cutter is necessary, as shown in Fig. 4-2. Here it shows the impossibility of cutting a pipe in a close-fitting space with a wheel-and-roller cutter. A 360° rotation of the entire cutter is necessary to cut the pipe. However, with a three-wheel cutter, a rotation of slightly more than 130° is all that is

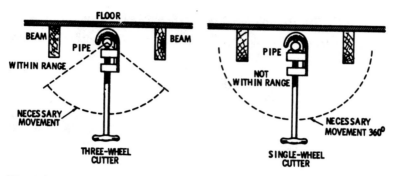

Fig. 4-2. The difference between a three-wheel and a single-wheel pipe cutter when used in a confined space.

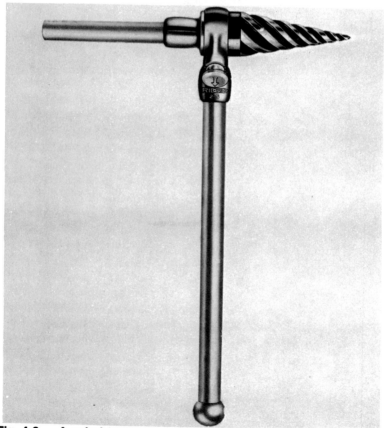

Fig. 4-3. A spiral reamer. *(Courtesy Ridge Tool Co.)*

necessary to cut the pipe. The cutter wheels on either type are easily removed and renewed when they become dull or nicked.

A little more care must be taken when starting a cut with a three-wheel cutter than with the roller type to make certain the cut is straight. In addition, the three-wheel type leaves more of an outside burr. This burr, as well as the burr on the inside of the pipe, must be removed on every cut to avoid future trouble with clogged pipes. The outside burr will also interfere with starting the

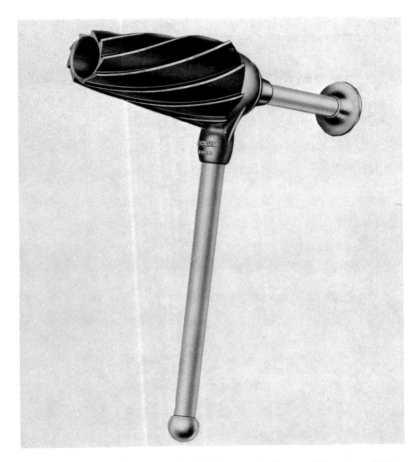

Fig. 4-4.   A spiral reamer for 2½" through 4" pipe. *(Courtesy Ridge Tool Co.)*

pipe threader, so it should be removed with a file or rasp before attempting the threading operation. The internal burr in pipe sizes from ⅛ in. through 2 in. I.D. can be easily removed by using a reamer of the type shown in Fig. 4-3. For pipe sizes 2½ in. through 4 in. I.D. a reamer of the type shown in Fig. 4-4 can be used.

Pipe Threading

Having cut the pipe to proper length, filed off the outer shoulder, and reamed out the burr, it is now ready for the threading operation. The threads may be cut on the pipe ends by means of:

1. Hand stock and dies.
2. Pipe-threading machines.

Hand-Operated Threaders

The hand stock and dies are portable, and are generally used for small jobs, especially for threading pipe of the smaller sizes. Threading machines are for use where a large amount of threading is done.

Fig. 4-5 shows a hand-operated stock and dies combining threading facilities for three different pipe sizes in one unit. This threader is for pipe size ⅜ in. through ¾ in. and is instantly ready

Fig. 4-5.   A three-way stock and dies for threading ⅜″, ½″, and ¾″ pipe. *(Courtesy Ridge Tool Co.)*

Fig. 4-6.   Ratchet-type pipe dies feature interchangeable dies plus ratchet action for ease of operation. *(Courtesy Ridge Tool Co.)*

to cut any one of three sizes of threads (⅜, ½, and ¾ in.). Each die set is securely locked in place and includes a guide to insure straight and true threads. The dies can be reversed in the holder for close-to-wall threading.

Another hand-operated threader (Fig. 4-6) has a ratchet handle and will accept dies for threading pipe from ¼ in. through 1¼ in. The die heads snap in from either side and push out easily for fast changing. The ratchet permits threading pipe in close quarters.

Power Threaders

Where large amounts of pipe are to be threaded, a power threading machine (Fig. 4-7) offers a great saving in time and labor. In addition, the quality of the thread is usually better than with a hand-operated

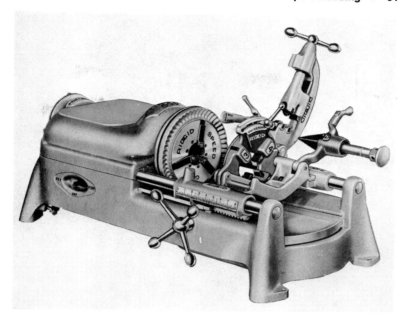

Fig. 4-7.   A power threading machine. This unit has dies to thread pipe from ⅛″ I.D. through 2″ I.D. *(Courtesy Ridge Tool Co.)*

unit. The most versatile, and usually most expensive, type of power threader is a self-contained unit powered by an electric motor. This threader has a built-in pipe cutter, a reamer, and a pump to circulate threading oil and direct it to the proper place where it floods the threading area. The dies for different sizes of pipes are instantly interchangeable, making the machine ideal for all types of plumbing and pipe fitting. A typical pipe die for use with a power threading machine is shown in Fig. 4-8.

The geared pipe threader shown in Fig. 4-9 can be powered by a power drive, Fig. 4-10, through a universal joint, or a handheld unit shown in Fig. 4-11. Gearing makes it possible to drive this threader by hand if electric power is not available. The threader shown in Fig. 4-9 is used on 2½-in. through 4-in. pipe. It is not necessary to change die segments when this threader is used; the same segments are used to thread 2½-in., 3-in., and 4-in. pipe.

It often happens that a thread in a fitting or a tapped (threaded)

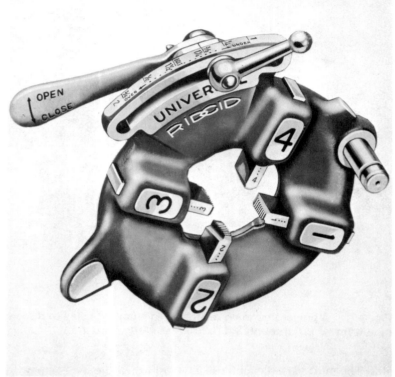

Fig. 4-8. Quick-opening dies used with the pipe machine shown in Fig. 4-7. *(Courtesy Ridge Tool Co.)*

opening in equipment, such as pumps, boilers, water heaters, etc., is damaged. Internal (female) threads can be re-cut with a pipe tap. Care must be taken to start the tap straight; it should be turned slowly and oiled liberally. If the existing thread is badly damaged it may be necessary to run the tap in two or three threads, then back it out to remove shavings, and start again.

Making Accurate Measurements

The plumber and pipe fitter use end-to-end and end-to-center measurements when installing piping. An end-to-end measurement is

Fig. 4-9. A geared pipe threader that can be used for power or hand threading. *(Courtesy Ridge Tool Co.)*

from the thread on the end of a pipe nipple or section of pipe to the end of the thread on the other end. This is shown in Fig. 4-13 as dimension X. An end-to-center measurement is from the end of a thread on a pipe nipple or section of pipe to the center of a fitting on the other end. This is shown as dimension X^1.

The space from the end of the threads in a fitting to the center of that fitting is called the "takeoff," shown as dimension Y. The make-up or end-of-threads point in a fitting is shown as Z. An end-to-center measurement can be made in either of two ways: (1) Allowing for the make-up of a fitting in place to be measured from, it is determined that a length of pipe 24 in., end-to-center of a fitting, is needed to reach a certain point. If the takeoff, Y, of the fitting is ½ in. the pipe would need to be 23½ in. long, end-to-end

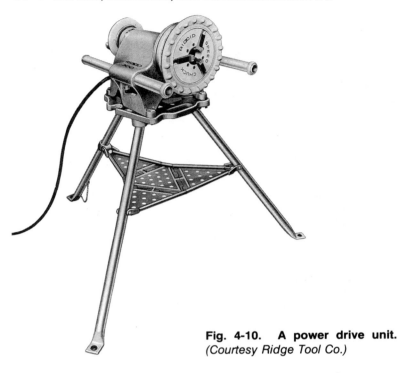

Fig. 4-10.   A power drive unit.
(Courtesy Ridge Tool Co.)

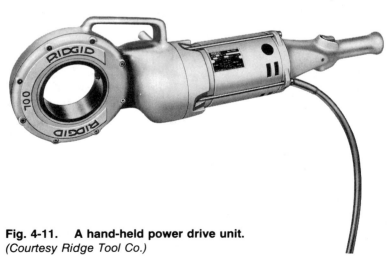

Fig. 4-11.   A hand-held power drive unit.
(Courtesy Ridge Tool Co.)

Table 4-1   Drill Sizes for Standard Pipe Taps*

Pipe Size	Threads per Inch	Drill Size
$\frac{1}{8}$	27	$\frac{11}{32}$
$\frac{1}{4}$	18	$\frac{7}{16}$
$\frac{3}{8}$	18	$\frac{19}{32}$
$\frac{1}{2}$	14	$\frac{23}{32}$
$\frac{3}{4}$	14	$\frac{15}{16}$
1	$11\frac{1}{2}$	$\frac{15}{32}$
$1\frac{1}{4}$	$11\frac{1}{2}$	$1\frac{1}{2}$
$1\frac{1}{2}$	$11\frac{1}{2}$	$1\frac{23}{32}$
2	$11\frac{1}{2}$	$2\frac{3}{16}$
$2\frac{1}{2}$	8	$2\frac{5}{8}$
3	8	$3\frac{1}{4}$
$3\frac{1}{2}$	8	$3\frac{3}{4}$
4	8	$4\frac{1}{4}$
$4\frac{1}{2}$	8	$4\frac{3}{4}$
5	8	$5\frac{5}{16}$
6	8	$6\frac{3}{8}$

*To secure the best results, the hole should be reamed before tapping with a reamer having a taper of $\frac{3}{4}$ " per foot.

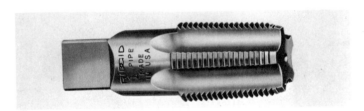

Fig. 4-12.   A pipe tap for cutting internal threads. *(Courtesy Ridge Tool Co.)*

of the threads; $23\frac{1}{2}$ in. plus $\frac{1}{2}$ in. equals 24 in. end-to-center. (2) A more accurate way to make the end-to-center measurement is to make the fitting on to a length of pipe first, then measure 24 in. from the center of the fitting, mark the pipe, and cut and thread it. This method is better because it eliminates the guesswork of allowing for the make-up of a thread.

Calculating Offsets

In pipe fitting, the term *offset* may be defined as *a change of direction (other than 90°) in a pipe that brings one part out of, but*

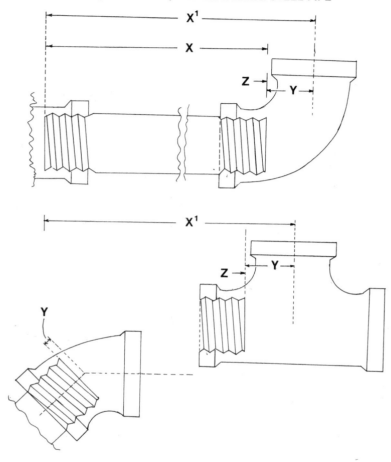

Fig. 4-13.   The key to accurate measurements.

parallel with, the line of another pipe. An example of this is illustrated in Fig. 4-14 where it is necessary to change the position of pipe line, L, to a parallel position, F, in order to avoid some obstruction, such as wall E. When the two lines, L and F, are to be fitted with elbows having an angle of other than 90°, the pipe fitter must find the length of the pipe H connecting the two elbows, A and C. The distance BC must also be determined in order to fix the point A, so that elbows A and C will be in alignment. There

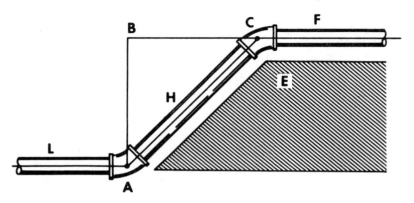

Fig. 4-14. A pipe line connected with 45° elbows showing offset and the method of finding the length of the connecting pipe.

are several methods of solving this problem; four of these methods are given here.

METHOD 1

In the triangle ABC

$$(AC)^2 = (AB)^2 + (BC)^2$$

from which

$$AC = \sqrt{(AB)^2 + (BC)^2}$$

Example—If the distance between pipe lines L and F in Fig. 4-14 is 20 inches (offset AB), what length of pipe H is required to connect with the 45° elbows A and C? When 45° elbows are used, both offsets are equal. Thus, substituting in the equation,

$$AC = \sqrt{20^2 + 20^2} = \sqrt{800} = 28.28 \text{ inches}$$

The length of the pipe just calculated does not allow for the projections of the elbows. This must be taken into account, as shown in Fig. 4-15. The projection of a fitting is the measurement between the end of the female thread in a fitting and the center of the fitting. Another and more commonly used term for projection is "takeoff."

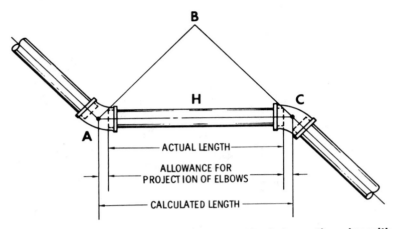

Fig. 4-15. The calculated and actual length of connecting pipe with elbows other than 90°.

METHOD 2

The following rule will be found convenient in determining the length of the pipe between 45° elbows.

Rule—*For each inch of offset, add* ⁵³/₁₂₈ *of an inch, and the result will be the lengths between centers of the elbows.*

Example—Calculate the length AC (Fig. 4-15) by preceding rule.

$$20 \times \frac{53}{128} = \frac{1060}{128} = 8\%_{32}$$

Adding this to the offset,

$$2 + 8\%_{32} = 28\%_{32} \text{ inches}$$

This is the calculated length; deduct the allowance for the projection of the elbows, as in Fig. 4-15.

METHOD 3

Elbows are available with angles other than 45° and 90°. For instance, angles of 60°, 30°, 22½°, 11¼°, and 5⅜° are manufactured

Table 4-2   Elbow Constants

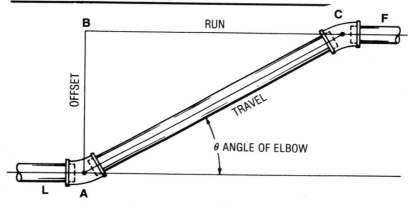

Elbow Angle	Elbow Centers (AC)	Run (BC)
60°	1.15	0.58
45°	1.41	1.00
30°	2.00	1.73
22½°	2.61	2.41
11¼°	5.12	5.02
5⅝°	10.20	10.15

and sometimes encountered by the pipe fitter. When such elbows are used, the distance between their centers can easily be found by using the constants listed in Table 4-2.

Rule—*To find the length between centers of the elbows, multiply the offset by the constant for the elbows used.*

Referring to the figure in Table 4-2

AC = offset AB × constant for AC

BC = offset AB × constant for BC

Example—If the distance between pipe lines L and F (offset AB) is 20 inches, what is the length of offset BC and the distance AC between the centers of the elbows, if 22½° elbows are used?

From the table, the constant for AB, for a 22½° elbow, is 2.41. Substituting the values in the proper equation,

$$BC = 20 \times 2.41 = 48.2 \text{ inches}$$

For the distance (travel) between the elbows, the constant in the table is 2.61. Substituting the values in the proper equation,

$$AC = 20 \times 2.61 = 52.2 \text{ inches.}$$

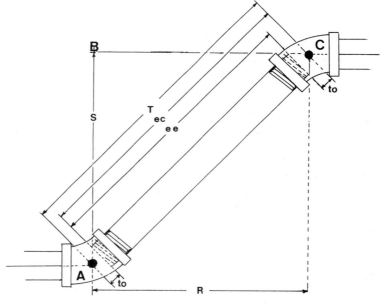

R run, the lineal space required for the offset.
S set (offset), the measurement of the offset.
T travel, the center-to-center measurement of the offset piping.
ec end-to-center measurement of the offset piping.
ee end-to-end measurement of the offset piping.
to takeoff, space from end of pipe to center of fitting.

Fig. 4-16.   The terms used for an offset.

METHOD 4

Method 4 involves the use of trigonometry. This method is explained in the Mathematics chapter of Vol. III. The natural trigonometric functions are the ones of value in ordinary calculations and should be thoroughly understood. They are used with the Table of Natural Trigonometric Functions in the Mathematics chapter mentioned above.

Three terms are used when calculating offsets:

R—is the *run* or developed length of an offset.

S—is the *set*, the offset measurement.

T—is the *travel*, the measurement between the *centers* of the angle fittings used in an offset.

The constants shown in Table 4-2 can be used to find the run, set, and travel of simple offsets.

After calculating an offset, using any method, one more step is needed to know the *length of pipe* for the offset. The travel is the distance between the *centers* of the fittings. The space (takeoff) between the *ends of the threads* in both angle fittings and the *centers* of the fittings must be subtracted from the travel in order to know the length of cut pipe (or nipple if 12 in. or less) required for the offset. The takeoff is shown in Fig. 4-16.

Table 4-3 shows the multipliers used to calculate offsets using 5⅝°, 11¼°, 22½°, 30°, 45°, and 60° fittings. The multipliers can be used for both elbow and wye fittings. Fig. 4-16 shows a 45° offset and explains how the terms used above are used.

Table 4-3 Multipliers Used to Calculate Offsets

To Find Side	When Known Side Is	Multiply Side	Using 5⅝° Elbows	Using 11¼° Elbows	Using 22½° Elbows	Using 30° Elbows	Using 45° Elbows	Using 60° Elbows
T	S	S	10.187	5.125	2.613	2.000	1.414	1.155
S	T	T	.098	.195	.383	.500	.707	.866
R	S	S	10.158	5.027	2.414	1.732	1.000	.577
S	R	R	.098	.198	.414	.577	1.000	1.732
T	R	R	1.004	1.019	1.082	1.155	1.414	2.000
R	T	T	.995	.980	.924	.866	.707	.500

R = run   S = set   T = travel

Example 1—A run of pipe must be offset 16″ in order to clear an obstruction. What is the *length of pipe* needed?

$$1.41 \text{ (constant for 45° offset)}$$
$$\times \quad 16$$
$$8\ 46$$
$$\underline{14\ 1\ }$$
$$22.56$$

22.56 is the travel, the *center* to *center* distance, between the 45° elbows. 22.56 in. is rounded off to the nearest ⅛″ or 22½″.

The takeoff of the size 45° elbow used in this example is ⅜″. ⅜ × 2 (both fittings) = ¾″. Subtracting ¾ from 22½ = 21¾. The length of pipe needed for this 16″ offset is 21¾″ end to end of pipe.

Example 2—A run of pipe must be offset 16″ using 60° elbows. What is the length of pipe needed?

$$1.15 \text{ (constant for 60° offset)}$$
$$\times \quad 16$$
$$6\ 90$$
$$\underline{11\ 5\ }$$
$$18.40$$

18.40 is rounded off to the nearest ⅛″ or 18½″. 18½″ is the travel or *center* to *center* distance between the two 60° elbows.

Table 4-4   Approximate Length of Usable Pipe Thread

Pipe	Distance Between Centers	Fittings	Center to Face of Fitting (Subtract)	Allowance for Threads (Add)	Overall Pipe Length
A	20″	2	1⁵⁄₁₆″	½″ + ½″	19¹¹⁄₁₆″
B	43″	2–3	1⁵⁄₁₆″ + 1⁵⁄₁₆″	½″ + ½″	41³⁄₈″
C	6″	3–4	1⁵⁄₁₆″ + 1⁵⁄₈″	½″ + ½″	4¹⁄₁₆″
D	24″	4–6	1⁵⁄₈″ + 1¹⁄₁₆″	½″ + ½″	22⁵⁄₁₆″
E	16″	5–6	1¹⁄₁₆″ + 1⁵⁄₁₆″	½″ + ½″	14⁵⁄₈″

The takeoff of this size 60° elbow is ½″. ½ × 2 (both fittings) = 1″. Subtracting 1″ from 18½″ = 17½″. The length of pipe needed for this 16″ offset is 17½″, end to end of pipe.

The secret of making tight joints may be summed up as follows:

1. The threads must be clean.
2. Good-quality pipe-joint compound or Teflon tape must be used.
3. The joint must not be tightened so rapidly as to appreciably change the temperature of the metal.

Are especially long threads favorable for tight joints?

Answer: No. The longer the thread, the greater the friction will be in the tightening process.

Pipe-Fitting Examples

For most plumbing or pipe-fitting installations, a sketch should be made showing all the fittings and lengths of pipe necessary to complete the job. This sketch need not be elaborate, but simply a freehand pencil drawing. Fig. 4-17 shows a typical sketch for part of a boiler installation.

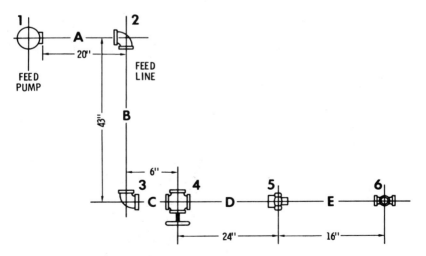

Fig. 4-17. A centerline sketch with dimensions for pipe fitting entirely by measurement.

Assembly

The pipe on large jobs is usually cut according to a sketch or working drawing and partly assembled at the shop. If no mistakes have been made in following the dimensions on the drawing, and if the drawings are correct, the pipe and fittings may be installed without difficulty—that is, the last joint will come together. The last joint must of necessity be a union, preferably either a ground-joint union or a flange union.

On small jobs, no sketch is necessary. The plumber proportions the pipe lengths mostly "by eye," taking occasional measurements where necessary during the progress of the work. It should be noted that, with the great variety of fittings available, any pipe system may be arranged in a number of different ways. The proper selection of these fittings, and the general arrangement of the system so that it will be direct, simple, accessible for repairs, etc., is an index of the pipe fitter's ability.

Joint Compound

In making up screwed joints, a good joint compound is mandatory. This compound serves two useful purposes—it lubricates the joint, making the tightening process much easier, and also forms a seal to insure a tight joint. Many different materials can be used as a joint compound, but the best is that manufactured for this purpose. Teflon tape, which is also available, is easy to use and works well and efficiently. If neither of these types of joint materials is available, white or red lead, or graphite, can be used with some success. Red lead will provide a tight joint, but has the disadvantage of making it difficult to unscrew the joint in case of future repairs.

In applying the joint compound, *it should be put on the male thread only.* If put on the female thread, some of the compound will lodge inside the pipe, forming an obstruction or contaminating the liquid, which will flow through the pipe.

Joint Make-up

When making up a joint, it should not be tightened too rapidly. The tightening process produces heat (because of friction), which may expand the pipe enough so that the proper number of turns cannot be made. After the joint has cooled, it may be loose enough to allow a leak. The approximate distances that different sizes of pipe should extend into fittings for a tight joint are listed on Table 4-5. This length of thread must be taken into account when cutting the pipe to fit in a given space.

In order to avoid frequent changes of dies, it is best to make all lines of one size, when possible, before making up lines having a different size of pipe. In the installation in Fig. 4-17, the lines are all ¾-in. pipe. The line in Fig. 4-17 consists of pipes A, B, C, D, and E, the pump connection 1, and fittings 2, 3, 4, 5, and 6. Determine the overall dimensions of A, B, C, D, and E by preparing a table, using the measurements from the free-hand sketch and the dimensions given for standard fittings in Table 4-5 below.

To determine the overall length of pipe A, notice that the distance between the face of the pump connection and the center of elbow 2 is 20 inches. This is shown in detail in Fig. 4-17. From Table 4-5, the distance from the center of a standard ¾-in. elbow

Table 4-5　Dimensions of Standard Malleable-Iron Fittings

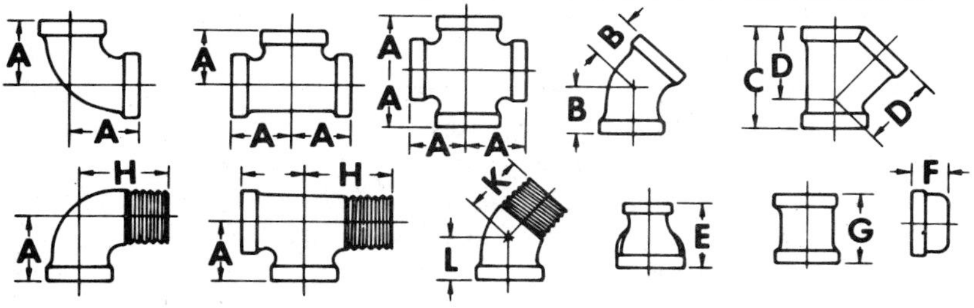

Size in.	⅛	¼	⅜	½	¾	1	1¼	1½	2	2½	3	3½	4	4½	5	6
A. in.	11/16	13/16	15/16	1⅛	1 5/16	1 7/16	1¾	1 15/16	2¼	2 11/16	3⅛	3 7/16	3¾	4 1/16	4 7/16	5⅛
B. in.		¾	13/16	⅞	1	1⅛	1 5/16	1 7/16	1 11/16	1 15/16	2 3/16	2⅜	2⅝	2 13/16	3 1/16	3 7/16
C. in.			2⅛	2½	2⅞	3 7/16	4 1/16	4½	5 7/16	6¼	7¼		8⅞			
D. in.			1 7/16	1 11/16	2	2 7/16	2 15/16	3 5/16	4 1/16	4 11/16	5 9/16		6 15/16			
E. in.		1	1⅛	1¼	1 7/16	1 11/16	2 1/16	2 5/16	2 13/16	3¼	3 11/16	4	4⅜			
F. in.	17/32	⅝	¾	⅞	1 1/16	1 3/16	1¼	1 5/16	1 7/16	1⅝	1¾	1 15/16	2		2 5/16	2 9/16
G. in.		1 1/16	1 3/16	1 5/16	1½	1 11/16	1 15/16	2⅛	2½	2⅞	3 3/16					
H. in.	1⅛	1 5/16	1 7/16	1⅝	1⅞	2⅛	2½	2 11/16	3 3/16	3 13/16	4½		5 11/16			
K. in.			15/16	1 1/16	1 3/16	1 5/16	1½	1 11/16	1⅞	2¼	3		3¾			
L. in.				⅝	11/16	13/16	15/16	1 1/16	1¼	1⅜	1 11/16		2⅛		2½	

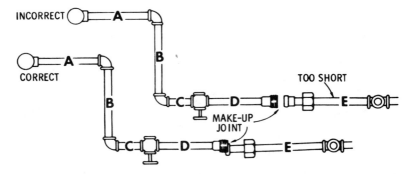

Fig. 4-18. The correct and incorrect make-up of the feedline shown in Fig. 4-17.

to its face is $^{15}\!/_{16}$ in. From Table 4-5, the length of the thread that will be screwed into the fitting is ½ in. for a ¾-in. pipe. Therefore, the total length of pipe A will be

$$20 - 1^{5}\!/_{16} + ½ + ½ = 19^{11}\!/_{16} \text{ inches}$$

The lengths for pipes B, C, D, and E are determined in a similar manner, the only difference being that the measurements for fittings 4 and 5 are not listed in Table 4-5. These measurements, which vary according to the manufacturer, must be taken on the actual valve and union to be installed.

Fig. 4-18 shows the results of poor and good workmanship in the final make-up where the line is joined with a union. If pipe E is made too short, a gap at the union will make it difficult to bring the make-up joint together. Even if brought together, the system will be under an undue strain. The proper dimensions will result in the make-up joint springing into position snugly, with no appreciable stress or strain on any part of the system.

Pipe Supports

All piping should be supported in accordance with approved standards. Horizontal runs may be supported by hangers fastened to the ceiling or by wall brackets. The supports should be placed at frequent intervals—in no case more than 12 ft. apart. It is more desirable to support the pipe along concrete or masonry walls rather than wood floor joints since the wood joints may allow and amplify vibrations in the lines.

Vertical pipe runs are more difficult to support. A shoulder clamp bearing on the floor slab through which the pipes pass may be used. An alternate method is to build a supporting platform under the lower elbows to bear the weight of the risers. The upper horizontal runs should be well supported so as not to add any additional weight to the vertical runs. In addition, the vertical pipes should be clamped to adjacent walls or columns to hold the pipes rigid. These clamps should not be expected to bear any weight.

Pipe Expansion

Care must be taken in the design and installation of steam or hot-water pipes to provide for variations in length and form due to temperature changes. If these variations are not adequately taken care of, the system will be subject to undue stress and strain, resulting in possible damage to valves, joints, and fittings. The pipes should be securely anchored at certain points, while at other points, sliding or flexible hangers must be used. Expansion and contraction of the pipes is taken care of by such a method of support and by the addition of large-radius bends or expansion joints. Some of the commercially available pipe bends are shown in Fig. 4-19, while their critical dimensions are listed in Table 4-6.

Corrosion

In the treatment or prevention of corrosion, special consideration must be given to local conditions. In all cases where iron or steel pipes are exposed to moist air, they should be protected by water-proof and durable coatings. Internal corrosion is caused by the solvent or oxidizing properties of water and accelerated by the salts and gases (including air) dissolved in it. This makes the purification and treatment of boiler feed water necessary. The safest plan in this case is to consult a competent chemist experienced in the analysis and treatment of boiler feed water, and follow his recommendations.

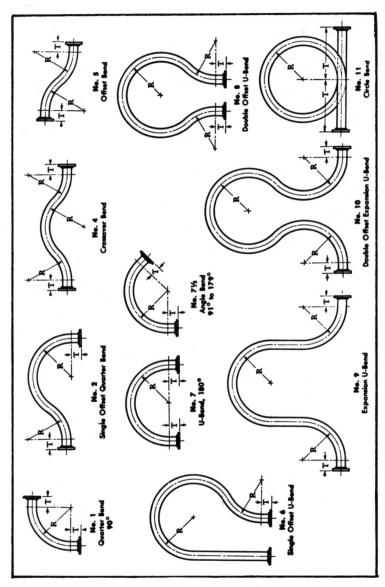

Fig. 4-19. Radii and tangents of standard pipe bends.

Table 4-6 Radii and Tangents of Standard Pipe Bends

Column 1 Weight of Thickness of Pipe	Column 2 Size of Pipe	Column 3* Minimum Recommended Radius		Column 4* Minimum Tangent Including Pull Length		Column 5* Minimum Tangent After Cutting for Cranelap Joints		Column 6 Minimum Tangent After Cutting for Welding Ends Threaded Ends, Screwed Flanges	
	Inches	Cold Bend	Hot Bend	Cold Bend	Hot Bend	Cold Bend	Hot Bend	Cold Bend	Hot Bend
Standard or Heavier Weight	¼	1¼	1¼	3	3	...	...	1	1
	⅜	1⅞	1⅞	3	3	...	...	1¼	1¼
	½	2½	2½	3	3	2	2	1½	1½
	¾	3¾	3¾	3	3	2	2	1¾	1¾
	1	5	5	3	3	2	2	2	2
	1¼	6¼	6¼	3	3	2½	2½	2	2
	1½	7½	7½	3	3	3	3	2½	2½
	2	10	10	3	3	4	4	3	3
Standard or Extra Strong Weight for Cold Bend Process and Standard or Heavier Weight for Hot Bend Process	2½	12½	12½	18	12	5	5	4	4
	3	15	15	18	12	6	6	4	4
	3½	17½	17½	18	12	6	6	5	5
	4	20	20	18	12	6	6	5	5
	5	30	25	18	12	7	7	6	6
	6	42	30	18	12	7	7	7	7
	8	...	40	...	18	...	9	...	9
	10	...	50	...	18	...	12	...	12
	12	...	60	...	24	...	14	...	14
	14 OD	...	70	...	24	...	16	...	16
⅜-inch Thick†	16 OD	...	96	...	30	...	18	...	18
	18 OD	...	108	...	30	...	18	...	18
½-inch or Thicker	16 OD	...	80	...	30	...	18	...	18
	18 OD	...	90	...	30	...	18	...	18
	†20 OD	...	100	...	36	...	20	...	18
	†24 OD	...	120	...	36	...	24	...	18

Courtesy Crane Co.

*Column 3 radii preferably should equal or exceed dimensions shown; increased flattening and wall thinning usually result from smaller radii. Column 4 dimensions are intended to cover bends that will be further fitted in the field. In Column 5, lapped stub ends welded on are recommended for sizes 2 inch and smaller, and for 18-, 20-, and 24-in. sizes.
†The table does not include sizes 20- and 24-in. of ⅜-in. wall thickness. Bending such pipe is not practical since "buckling" and excessive "wrinkles" will develop in the crotch section.

Pipe Sizes

To efficiently convey water or steam through a pipe under pressure, the pipe must not be too small or there will be an undue drop in pressure, resulting in insufficient flow. If the pipe is too large, the initial cost is impractical, and in the case of steam, an increase of condensation in the pipe renders the system inefficient.

CHAPTER 5

Blueprints and Elevations

Reading and understanding blueprints are a necessary prerequisite for roughing-in plumbing and heating piping. This is especially true when working on commercial, industrial, and public buildings. Architects and engineers demand high-quality workmanship. The finished product shows the ability of the workmen to read and understand the prints. Before the work shown in Chapter 8, Roughing-In Plumbing, can be performed, the groundwork, figuring out the necessary measurements, must be done.

Building plans are printed in several different forms. The most common are blueprints, which are white lines on a blue background, and blue line drawings—blue lines on a white background. Architects and engineers have their individual style in preparing plans, but the basic principles are the same.

A building that starts out as an idea becomes a reality partly because the architect and the mechanical engineer know how to translate ideas into lines and symbols called *blueprints*. The architect and the engineer deserve only part of the credit for the finished project. It takes skilled mechanics, the building tradesmen, among whom are the plumber and the pipe fitter, to translate the language of blueprints into the finished project.

Two types of plans are used on most building projects: architectural plans and mechanical plans. Architectural plans show footing plans, floor plans, roof plans, details such as framing, and room finish schedules, which the plumber and pipe fitter should be able to read in order to fit the building piping into the available spaces. Architectural plans are more rigid than mechanical plans; for instance, a footing must be placed exactly where the plans indicate, or the building may end up on someone else's property.

Mechanical plans may indicate that piping originates at a definite point, but unforeseen obstacles may cause it to deviate in some degree from the route shown on the plans. These obstacles are called "job conditions," and the plumber and the pipe fitter must be able to read and interpret the plans in order to overcome the job conditions and get the project finished in a workable manner.

Blueprints show room sizes, wall sizes, equipment layouts, pipe chases, room finish schedules, door sizes, and so on. It is often necessary to jump back and forth between different sheets to find information because it is impossible to show all the details on one sheet.

In addition to floor and site plans which must show large areas on a single sheet, most blueprints include detail sheets. A floor plan may show a toilet room location but it is impossible to show the details of the construction of this room using a scale of ⅛ in. = 1 ft. (⅛″ = 1′0″). The details are usually shown on a separate plan sheet using a much larger scale. The symbol for referring to a detail drawing will show the detail number and the plan sheet on which it is shown. The symbol may be a circle, square, triangle, etc., within which is placed the number and sheet. Fig. 5-3 shows a toilet room equipment layout and a detail symbol showing a marble shelf over the lavatories. The plumber would refer to this detail before roughing-in the piping for the lavatories.

Mechanical plans show the routing of pipes, ducts, electrical conduits, etc., but the exact locations and elevations can only be decided after checking the architectural plans.

The reproductions of actual blueprints shown in Figs. 5-1 through 5-4 show only the measurements needed to explain the basics of blueprint reading.

The following examples show how the plumber and pipe fitter, using the information from the different sections of the plans, find the exact points at which the fixtures should be roughed-in.

Figs. 5-1 and 5-3 are plans of the same room; Fig. 5-1 shows all the measurements needed to build the room. Fig. 5-3 shows

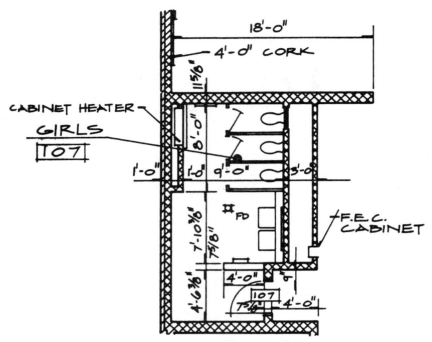

Fig. 5-1.   Floor plan of Girls' Toilet Room 107.

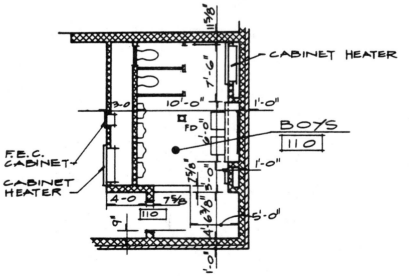

Fig. 5-2.   Floor plan of Boys' Toilet Room 110.

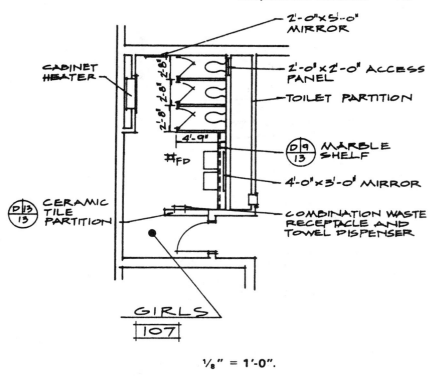

2'-0"x5'-0" MIRROR

CABINET HEATER

2'-0'x2'-0' ACCESS PANEL

TOILET PARTITION

4'-9"

FD

D 9
13 MARBLE SHELF

4'-0'x3'-0' MIRROR

D 13
13 CERAMIC TILE PARTITION

COMBINATION WASTE RECEPTACLE AND TOWEL DISPENSER

GIRLS

107

⅛" = 1'-0".

Fig. 5-3. Equipment layout Girls' Toilet Room 107.

where the equipment, toilets, lavatories, and cabinet heater will be placed. The cabinet heater centers in an 8'0" space, the water closets are enclosed by toilet partitions placed on 2'8" (32") centers, therefore the water closets will also be set on 2'8" centers (16" + 32" + 32" + 32" + 16").

The lavatories are centered in a 7'10⅜" space as shown in Fig. 5-1. Fig. 5-3 shows a detail XXX (detail 9 on plan sheet 13) of a marble shelf above the lavatories. This detail must be examined because the framing for the shelf could affect the roughing-in for these fixtures.

The next step is to determine the location of the fixtures in Boys' Toilet Room 110. Fig. 5-2 shows the room to be 21'0⅜" long (inside measurement).

$$4'6\tfrac{3}{8}''$$
$$+ \ 3'0 \ ''$$
$$+ \ 6'0 \ ''$$
$$\underline{+ \ 7'6 \ ''}$$
$$21'0\tfrac{3}{8}''$$

The lavatories will be centered from a point 10'6" from the 11⅝" wall.

$$7'6''$$
$$\underline{+ \ 3'0''} \quad (\text{½ of } 6'0'' \text{ space})$$
$$10'6''$$

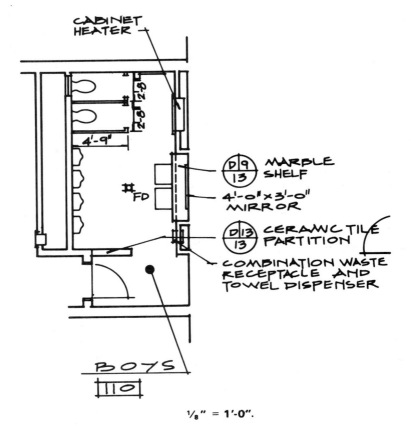

¼" = 1'-0".

Fig. 5-4. Equipment layout Boys' Toilet Room 110.

Fig. 5-4 shows the toilet partitions are on 2'8" (32") centers; therefore the toilets will be roughed-in on 32" centers (16" + 32" + 32" + 16").

The measurements for locating the four urinals are not shown and must be worked out as follows: The toilet room is 21'0⅜" long. Starting at the inside of the 11⅝" wall the space for the urinals is 10'6⅜" long, determined as follows:

$$
\begin{array}{rl}
& 7'6'' \\
+ & 6'0'' \\
+ & 3'0'' \\
\hline
& 16'6'' \\
- & 5'4'' \quad \text{(width of toilet partitions)} \\
\hline
& 11'2'' \\
- & 0'7\frac{5}{8}'' \quad \text{(width of entrance wall)} \\
\hline
& 10'6\frac{3}{8}'' \\
\end{array}
$$

Assuming that the "as built" space is 10'6" this space would be divided in this way, four urinals on 31½" centers with 15½" on each end.

Abbreviations are shown on blueprints wherever possible. F.E.C. on Fig. 5-1 indicates the location of a Fire Extinguisher Cabinet. Another common abbreviation is F.H.C., Fire Hose Cabinet. Common symbols used to identify pipe fittings and valves are shown in Figs. 5-5, 5-6, and 5-7.

Isometric Drawings

Fig. 5-8 is a section of a mechanical plan showing a men's and a women's toilet room. The rooms are back to back, and the piping for these rooms is concealed in a pipe chase between the two rooms. The plans indicate that soil, waste, vent, and hot and cold water piping are needed for these fixtures but it is up to the plumber to figure out how to rough-in the necessary piping. Fig. 5-9 is an isometric drawing of the piping shown in Fig. 5-8. Isometric drawings show what objects would look like viewed from above and slightly to one side. A good isometric drawing will show almost every fitting needed to rough-in the piping. Fig. 5-9 (A) shows the soil, waste, and vent piping. Fig. 5-9 (B) shows the cold water piping. Fig. 5-9 (C) shows the hot water piping. Mechanical engineers usually make isometric drawings and include them in the blueprints.

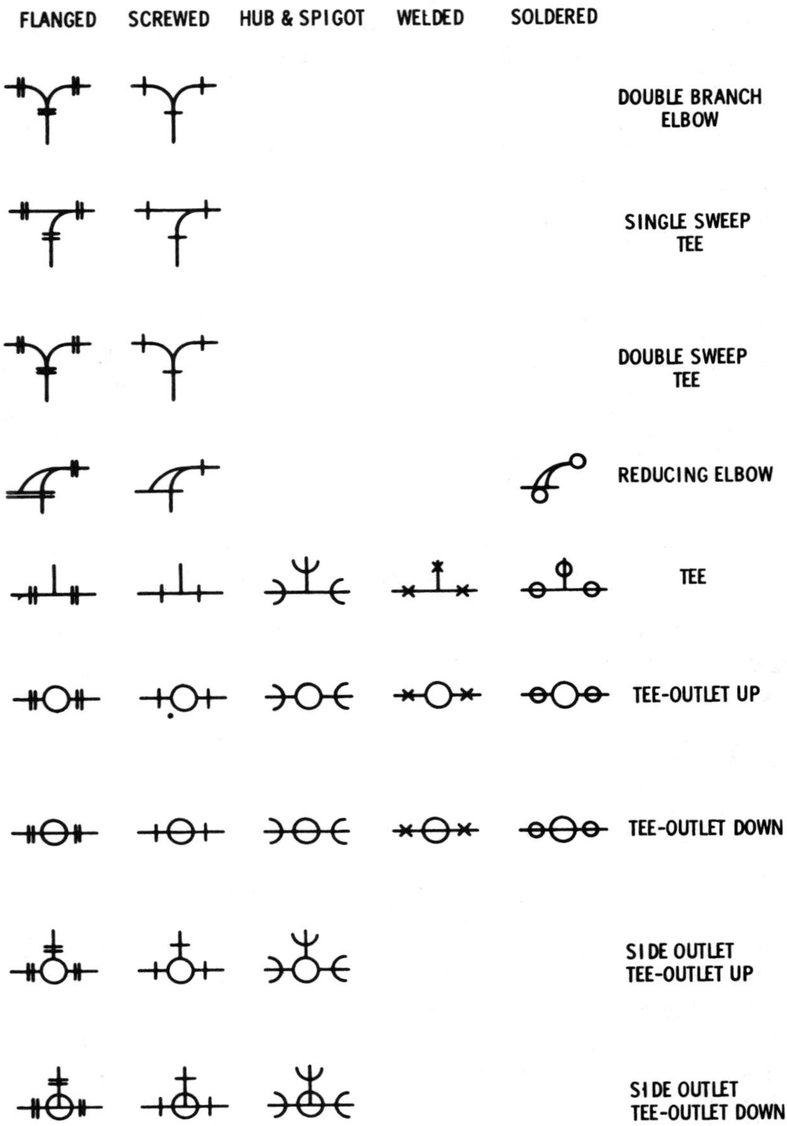

FLANGED	SCREWED	HUB & SPIGOT	WELDED	SOLDERED	
					DOUBLE BRANCH ELBOW
					SINGLE SWEEP TEE
					DOUBLE SWEEP TEE
					REDUCING ELBOW
					TEE
					TEE-OUTLET UP
					TEE-OUTLET DOWN
					SIDE OUTLET TEE-OUTLET UP
					SIDE OUTLET TEE-OUTLET DOWN

Fig. 5-5.   Symbols for pipe fittings and valves.

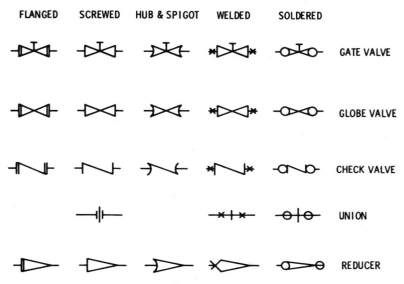

Fig. 5-6. Symbols for pipe fittings and valves (continued).

Elevations

In order to read and understand blueprints it is necessary to have a working knowledge of elevations. The elevations of various parts of a building are shown on the plans. The *bench mark* is the starting point for working out the different elevations. A bench mark may be an iron rod driven into the ground in a protected location; it may be a nail driven into a tree; however the bench mark is established, it will be used as the reference point for elevations until some permanent point such as the top of a footing or a finished floor is available for a reference point. Any number can be used to indicate the different elevations of a building project; quite often the number used is the approximate number of feet above sea level of the building site. A building site near the seacoast, at or near sea level, may use the number 100.00 as a reference point for elevations; while the number 750.00 might be used on land at that approximate height above sea level. In any case, the actual number used is not important; elevations higher than the bench mark are indicated by a higher number. If the finished first-floor elevation is shown as 750.00 and

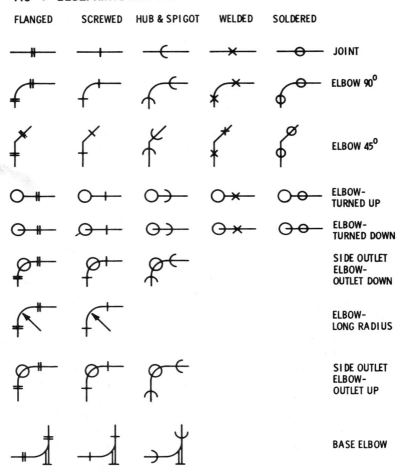

Fig. 5-7. Symbols for pipe fittings and valves (continued).

the second floor is shown as 760.00, it shows that the second floor is 10 ft. above the first floor. If the basement elevation is shown as 738.00, the basement will be 12 ft. lower than the finished first floor. The important thing to remember when working with elevations is that a number larger than the bench mark is *above* the bench mark; a smaller or lower number indicates a point *below* the bench mark.

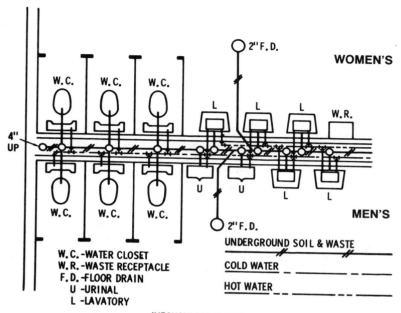

(MECHANICAL PLANS)

Fig. 5-8. A section of a mechanical plan showing a Men's & Women's Toilet Room.

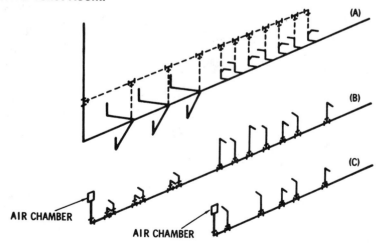

Fig. 5-9. Isometric drawing of the piping shown in Fig. 5-8.

ROOF ELEVATION _____ __780.00__

3RD. FLOOR ELEV. _____ __770.00__

2ND. FLOOR ELEV. _____ __760.00__

1ST. FLOOR ELEV. _____ __750.00__

BENCH MARK 747.50

BASEMENT ELEV. _____ __738.00__

Fig. 5-10.   A typical example of elevations.

As shown in Fig. 5-10, the bench mark is 747.50. The finished first-floor elevation is 750.00. The first floor is 2.50 (2½) ft. *above* the bench mark. The basement floor at an elevation of 738.00 is 9½ ft. (9.50) *below* the bench mark and 12 ft. below the finished first floor.

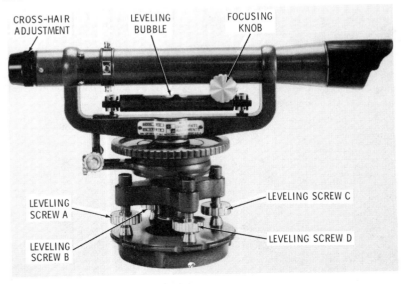

CROSS-HAIR
ADJUSTMENT

LEVELING
BUBBLE

FOCUSING
KNOB

LEVELING
SCREW A

LEVELING SCREW C

LEVELING
SCREW B

LEVELING SCREW D

Fig. 5-11.   An instrument level can be used to establish elevations.
(Courtesy David White Instruments)

The plumber and pipe fitter should be familiar with the use of the instrument level (Fig. 5-11). An instrument level is a telescope, mounted on a base equipped with leveling screws; the base is screwed onto a tripod (Fig. 5-12) when the instrument is in use. The instrument level can be rotated a full 360° horizontally, and when properly adjusted the level-indicating bubble will show that the instrument is level at any point of the compass. To set up the instrument the tripod should be opened and the legs spread apart so that the top of the tripod is approximately level with the legs of the tripod pressed firmly into the ground. The instrument level is then screwed onto the tripod and swung to a position lined up directly over adjusting screws A and C. Both screws should be turned equally and simultaneously, as shown in Fig. 5-13, to center the leveling bubble. Turning both screws "in" moves the bubble to the right. Turning both screws "out" moves the bubble to the left. When the bubble

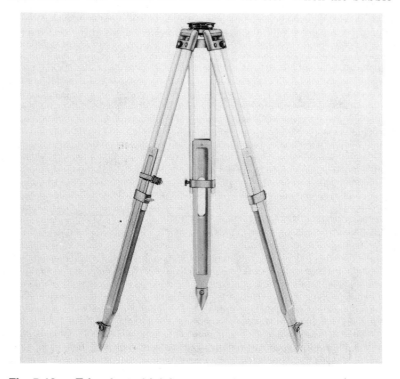

Fig. 5-12.   Tripod on which instrument level can be mounted. *(Courtesy David White Instruments)*

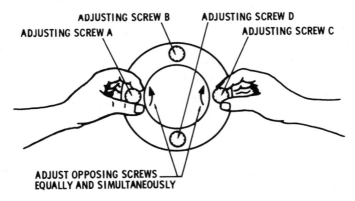

Fig. 5-13.   Leveling adjustments.

is centered, swing the instrument to line up directly over screws B and D and repeat the process. When the bubble is centered, return the instrument to line up over A and C again and readjust the screws to again level the instrument. The leveling process may have to be repeated several times until the instrument is in level position when

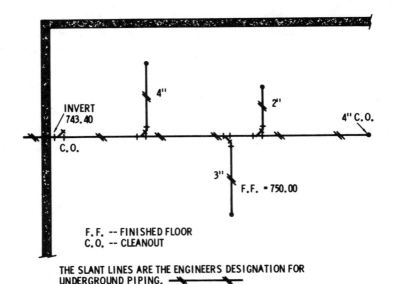

Fig. 5-14.   A partial mechanical plan of a building.

pointed in any direction. A word of caution: Do not *overtighten* the adjusting screws. Overtightening can damage the instrument.

Using an Instrument Level

Fig. 5-14 shows a partial mechanical plan of a building. The invert (or bottom inside) of the building drain is shown as 743.40. Although the finished floor elevation is shown as 750.00, this figure cannot

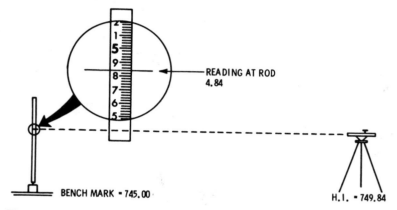

Fig. 5-15. Determining the height of an instrument.

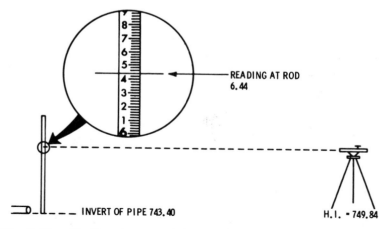

Fig. 5-16. Finding the building drain elevation.

be used to find the invert elevation because the construction is just starting. The plumber must set up an instrument level and set the building drain to the invert shown.

As shown in Fig. 5-15, with the rod held on the bench mark, a reading of 4.84 will be made. The bench mark is at 745.00; this figure is added to the 4.84 reading, to establish the H.I. (height of instrument). This figure, 749.84, is the actual elevation of the instrument. If the invert elevation is subtracted from the H.I., the result, 6.44, is the number the plumber should read when the building drain is at the correct elevation, as shown in Fig. 5-16:

$$749.84$$
$$-743.40$$
$$6.44$$

Fig. 5-17. A close-up view of a typical engineer's rod.

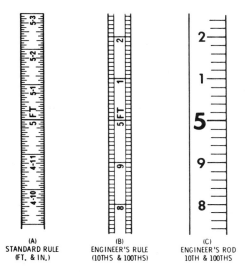

(A)	(B)	(C)
STANDARD RULE	ENGINEER'S RULE	ENGINEER'S ROD
(FT. & IN.)	(10THS & 100THS)	10TH & 100THS

Fig. 5-18. A standard rule, an engineer's rule, and an engineer's rod.

HELPFUL CONVERSIONS AND EQUIVALENTS

Approximate Conversions from Metric Measures

Symbol	When You Know	Multiply by	To find	Symbol
		LENGTH		
mm	millimeters	0.04	inches	in
cm	centimeters	0.4	inches	in
m	meters	3.3	feet	ft
m	meters	1.1	yards	yd
km	kilometers	0.6	miles	mi

Approximate Conversions to Metric Measures

Symbol	When You Know	Multiply by	To Find	Symbol
		LENGTH		
in	inches	*2.5	centimeters	cm
ft	feet	30	centimeters	cm
yd	yards	0.9	meters	m
mi	miles	1.6	kilometers	km

Fractional Inches	1/64	1/32	1/16	1/8	1/4	1/2	3/4
Decimal Inches	.016	.031	.063	.125	.25	.50	.75

Fig. 5-19. A metric standard conversion guide. *(Courtesy David White Instruments)*

It is very important when holding a rod to obtain a reading that the rod be held straight. If the rod is leaning toward the instrument, the reading will be high; if leaning away, the reading will be low. Fig. 5-17 shows a close-up view of the rod.

An engineer's 6-ft. rule shows feet and inches on one side and measurements in 10ths and 100ths on the other side. As shown in Fig. 5-18A, using an engineer's rule a measurement of 4'10⅞" can be converted to 4.91 by reading the 4'10⅞" on one side and turning the rule over and reading the 4.91 on the other side, Fig. 5-18B. This type of rule is very useful when working with prints which give some elevations or measurements in feet and inches and other measurements in feet, 10ths, and 100ths. Conversion guides such as Fig. 5-19 are helpful in converting measurements to or from metric measures.

CHAPTER 6

Soil, Waste, and Vent Piping

The soil, waste, and vent piping of a building is divided into several sections; building drain, soil and waste stacks, stack vents, vent stacks, and vent piping. A review of each part of the system will familiarize the reader with terms and definitions used throughout this three-book series. The definitions, shown in italics, are taken from the UNIFORM PLUMBING CODE and are often used as questions in written and/or oral examinations for Journeyman and Master plumbers' licenses.

Fig. 6-1 shows the soil, waste, and vent piping in a two-story residence. The building drain enters the building at the right side and receives the discharge from the two soil and waste stacks and from fixtures on the ground floor.

The building drain is that part of the lowest piping of a drainage system which receives the discharge from soil, waste, and other drainage pipes inside the walls of a building and conveys it to the building sewer beginning two (2) feet outside the building wall.

Two soil and waste stacks, one on each side of the building, extend up through the second floor, and as stack vents, continue through the roof. Each soil and waste stack receives the discharge from a water closet, lavatory, and bathtub on the second floor. The water closet and bathtub on the second floor, left side, are vented by a "wet" vent.

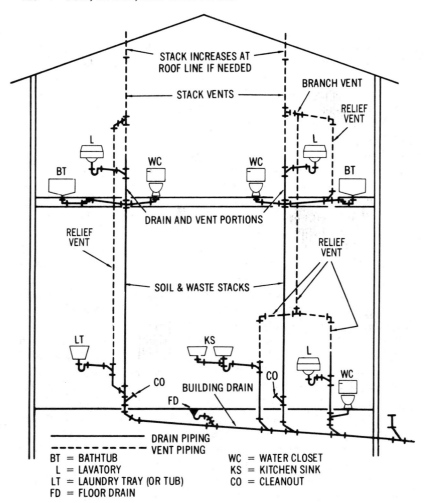

Fig. 6-1. Soil, waste, and vent piping in a house.

A wet vent is a vent which also serves as a drain. That portion of the stack above the lavatory waste inlet is called the "stack vent." *A stack vent is the extension of a soil or waste stack above the highest horizontal drain connected to the stack.* The bathtub shown on the right side, Fig. 6-1, second floor, requires a "relief" vent. The developed (overall) length of the drainage arm from the crown weir of the bathtub trap to the drain connection at the stack is more than 5 feet, whereas the maximum permitted distance for this measurement is 5 feet. Thus, a relief vent is required. (Local codes may vary in this regard.)

A vent system is a pipe or pipes installed to provide a circulation of air to or from a drainage system or to provide a circulation of air within such a system to protect trap seals from siphonage and back pressure.

A relief vent is a vent, the primary function of which is to provide circulation of air between drainage and vent systems or to act as an auxiliary vent on a specially designed system.

The laundry tray, on the left side, ground level, has a waste opening draining into the soil and waste stack. A relief vent extends upward from this fixture drain and connects to the stack vent above the second floor.

A floor drain is shown on the ground floor in Fig. 6-1. Although a floor drain is classed as a fixture it does not require a vent. A floor drain need not be vented and shall not be connected to drainage lines within five (5) feet of a stack or vent. Floor drain trap seals shall not be less than four (4) inches. The trap seal is the maximum vertical depth of liquid that a trap will retain, measured between the crown weir and the top of the dip of the trap.

Cleanout fittings are shown at the base of the soil and waste stacks. A cleanout is also shown outside the building wall where the building drain exits the building. Cleanouts shall be placed inside the building near the connection between the building drain and the building sewer or installed outside the building at the lower end of a building drain and extended to grade.

Proper venting prevents siphoning of traps by maintaining atmospheric pressure throughout the soil and waste piping and preventing the temporary formation of a vacuum in the drainage piping. Proper venting also allows any back pressure from the drainage system to be relieved through the vent piping and so protects the trap seals.

Batteries of water closets and lavatories are shown in Fig. 6-2. All are connected by drainage piping to the soil and waste stack. The water closets on the top floor (A) are vented by a loop vent. A loop vent is a vent connecting a horizontal branch or fixture drain with the stack vent of the originating waste or soil stack.

The lavatories, Fig. 6-2(B), and the water closets, Fig. 6-2(C),

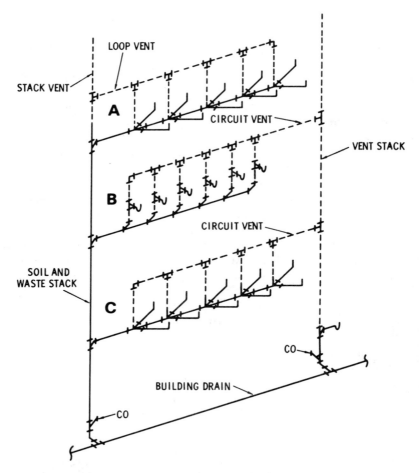

Fig. 6-2.   The difference between circuit and loop vents.

are vented by circuit vents. *A circuit vent is a branch vent that serves two or more traps and extends from in front of the last fixture connection of a horizontal branch to the vent stack. A vent stack is a vertical vent pipe installed primarily for the purpose of providing circulation of air to and from any part of the drainage system.*

A lavatory is shown at the base of the vent stack in Fig. 6-2. It is desirable to discharge the waste from a fixture such as a lavatory into or near the base of a vent stack in order to wash the vent stack.

INLET

WATER LEVEL IN TRAP

GREASES, OILS, ETC, FLOAT
TO THE TOP

WATER
OUT

OUTLET

CAST IRON, CONCRETE
VITREOUS TILE OR
FIBERGLASS BASIN

GRIT AND SAND SETTLE
TO BOTTOM

Fig. 6-3. A garage type floor drain.

Traps

All traps must have a water seal of not less than 2 inches and not more than 4 inches. Special installations requiring a seal of more than 4 inches must be approved by the local authority. One type of special installation, a garage type floor drain, is shown in Fig. 6-3. Greases, oils and gasoline, or other inflammable material float to the top where they can be skimmed off. This type of drain is designed to permit only water from near the bottom to enter the drainage piping. Most plumbing codes prohibit the use of certain types of traps.

No form of trap which depends for its seal upon the action of movable parts or concealed interior partitions shall be used. Full S traps are prohibited. Bell traps are prohibited. Crown vented traps are prohibited. No fixture shall be double trapped.

A plaster trap is shown in Fig. 6-4. Plaster traps are used in laboratories to catch materials that would clog the drainage piping. The water seal in the trap prevents the entrance of sewer gas. A "P" trap is shown in Fig. 6-5 and an "S" trap in Fig. 6-6.
Drum traps are permitted only for special conditions or for use on fixtures which are specially designed for drum traps. Where drum traps are permitted they must be vented.

Note

"May" is a permissive term.
"Shall" is a mandatory term.

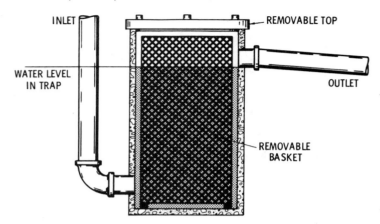

INLET

REMOVABLE TOP

WATER LEVEL
IN TRAP

OUTLET

REMOVABLE
BASKET

Fig. 6-4.   A plaster trap.

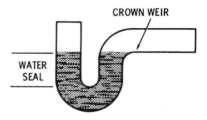

CROWN WEIR

WATER
SEAL

Fig. 6-5.   A "P" trap. Traps shall have a water seal of not less than 2 inches or more than 4 inches.

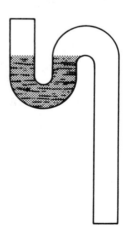

Fig. 6-6.   An "S" trap. "S" traps are prohibited by most plumbing codes.

Siphonage

In order to understand the operation of plumbing traps, one should first consider the principle of siphonage. The element of siphonage is a useful one in water closet traps, but undesirable in fixture traps and provision must be made to prevent fixture trap siphonage. A siphon is defined as "a bent tube or pipe, with legs of unequal length, used for drawing liquid out of a vessel by causing the liquid to rise within the tube, and over the rim or top."

The effect of siphonage on unvented traps is shown in Fig. 6-7. Part A shows the normal state of the trap—full of water with the bowl empty. Part B shows the bowl full of water with an air space between the stopper and the trap water. In part C the stopper has been removed and the water is beginning to flow from the bowl and the air space is being compressed. Part D shows the water flowing through the trap and into the drainage piping. In part E the bowl and one leg of the trap are empty, with water still flowing out of the trap due to momentum and siphonage. In part F water is being siphoned out of the other leg of the trap but the end surface of the water is breaking and dropping back to the bottom of the trap due to gravity. A solid column of water in the right leg of the trap

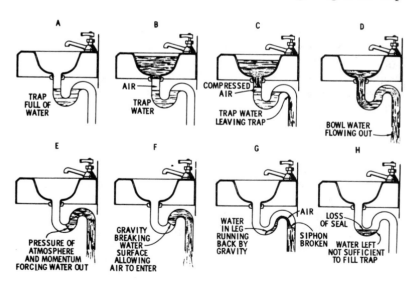

Fig. 6-7.   The effect of siphonage on an S trap.

in part G is broken by gravity, allowing air or sewer gases to enter and break the siphon. In part H the water remaining in the trap after the siphon is broken is not enough to retain the seal.

S traps have been outlawed in most areas because of the impossibility of proper venting of an S trap. The primary purpose of any plumbing trap is to provide and maintain a water seal which prevents sewer gases from entering a building.

Siphonage can also be caused by a flow of water passing a drainage inlet. Fig. 6-1 shows a laundry tray connected near the base of the soil and waste stack. Flushing the toilet connected to this stack on the second floor would cause a surge of water and waste material to drop down the stack. As this surge passed the laundry drain opening it would create a suction effect which could siphon the water out of the trap. The relief vent prevents siphoning; the suction effect would be broken by the air furnished by the relief vent.

Sizing of Soil, Waste, and Vent Piping

The sizing of soil, waste, and vent piping is determined by the load or number of fixture units discharging into the piping. A fixture unit is defined as *a quantity in terms of which the load-producing effects on the plumbing system of different kinds of plumbing fixtures are expressed on some arbitrarily chosen scale.* Or to put it in more simple terms, the sizing of soil, waste, and vent piping is determined by the quantity or volume of water and waste products discharged into the piping at any one time. Although different areas use different codes and some codes are more stringent than others in some respects, all codes assign basically the same values to fixtures in terms of fixture units.

Table 6-1 shows a partial listing of fixtures, fixture units, and trap sizes as prescribed by a widely used plumbing code. It shows the fixture unit load rating and the minimum trap size for the fixture. These fixture unit values apply only to areas using this code and are presented only to explain how the number of fixture units governs the size of soil, waste, and vent piping.

When a piping system is being designed, the trap size of a fixture will govern the size of the arm or branch drain into which the trap discharges. The drainage piping must never be smaller than the pipe size of the trap draining into the piping.

Table 6-1 Fixture Units

Kind of Fixture	Fixture Units	Minimum Trap & Trap Arm Size
Water Closet, tank type	4	3″
Water Closet, flush valve type	6	3″
Urinals, Wall Hung (2″ min. waste)	2	1½″
Urinals, pedestal	6	3″
Sinks, service	3	2″
Sinks, laundry	2	1½″
Bathtubs, with/without shower	2	1½″
Lavatory	1	1¼″
Sinks, residential (2″ min. waste)	2	1½″
Sinks, commercial, industrial, schools, etc.	3	1½″

Fixture traps are designed to be self-scouring; the trap size must not be increased to a point where the fixture discharge may be inadequate to maintain the self-scouring property. Every plumbing code contains tables showing the maximum number of fixture units which can be connected to any portion of the building drain or building sewer. The slope, or fall, in inches or fractions thereof per ft. also governs the number of fixture units connected to the soil, waste, and vent piping. Plumbing codes also contain tables

Table 6-2 Maximum Unit Loading and Maximum Length of Drainage and Vent Piping

Size of Pipe (inches)	1¼″	1½″	2″	2½″	3″	4″
Maximum Units						
Drainage Piping[1]						
Vertical	1	2[2]	16[3]	32[3]	48[4]	256
Horizontal[5]	1	2	8[3]	14[3]	35[4]	216
Maximum Lengths (feet)						
Vertical	45	65	85	148	212	300
Horizontal (unlimited)						
Vent Piping: Horizontal and Vertical						
Maximum Units	1	8	24	48	84	256
Maximum Lengths	45	60	120	180	212	300

[1]Excluding Trap Arm.
[2]Except Urinals and Sinks.
[3]Except Six-Unit Traps or Water Closets.
[4]Only four water closets or six-unit traps allowed on any vertical pipe or stack and not to exceed three water closets or six-unit traps on any horizontal branch or drain.
[5]Based on ¼ in. per ft. slope. For ⅛ in. per ft. slope multiply horizontal fixture units by a factor of 0.8.

showing the size and maximum permitted lengths of drainage and vent piping.

Table 6-1 shows only a partial listing of fixtures, fixture units, and trap sizes contained in this code. For information on fixture units in a specific area, the code in use in that area must be consulted.

Table 6-2 is taken from a widely used Plumbing Code. It is only a partial listing of the Maximum Loading and Maximum Length Table contained in the code book and is presented to explain to the reader how to use a Table to design pipe sizes and lengths. (The Table as shown above is not to be used in design work; the Code adopted in a specific area must be used in that area.)

Drainage and Vent Piping for Island Fixtures

Sinks or similar fixtures installed in island counter tops require a special method of venting. The UNIFORM PLUMBING CODE specifies that these fixtures be vented as follows:

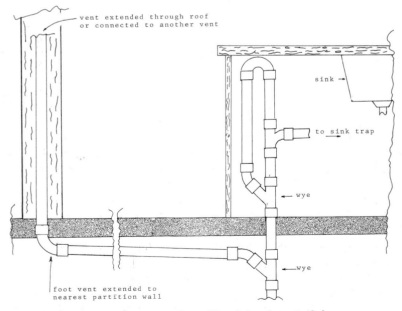

Fig. 6-8.   Correct method of installing island vent piping.

Traps for island fixtures and similar equipment shall be roughed-in above the floor and may be vented by extending the vent as high as possible, but not less than drainboard (counter top*) height and then returning it downward and connecting it to the horizontal sink drain immediately downstream from the vertical sink drain. The returned vent shall be connected to the horizontal sink drain through a wye branch fitting and shall, in addition, be provided with a foot vent taken off the vertical fixture vent by means of a wye branch immediately below the floor, extending it to the nearest partition and thence through the roof to the open air or may be connected to other vents at a point not less than six (6) inches above the flood rim level of the fixtures served. Drainage fittings shall be used on all parts of the vent below the floor level and a minimum slope of one quarter (¼) in. per ft. shall be maintained. The return bend used under the drainboard (counter top) shall be a one-piece fitting or an assembly of a forty-five (45) degree fitting, a ninety (90) degree fitting and a forty-five (45) degree fitting, in the order named. Pipe sizing shall be as otherwise required in this code.

Fig. 6-8 shows island venting. Island fixtures are classed under the heading of special wastes. The above Code applies only to areas which have adopted this code. The Code or Regulation in use in a specific area governs installations in that area.

Drainage Fittings

Fittings designed for use in drainage systems carrying liquid or water-carried wastes differ from standard I.P.S. (iron pipe size) pressure fittings. The fittings shown in Fig. 6-9 explain this difference.

Fig. 6-9 (A) shows a threaded drainage fitting, called a "Durham" fitting. This fitting is made with a recessed thread. When the fitting is screwed onto the pipe, the interior passageway is smooth, with no shoulder projecting. A pressure type I.P.S. fitting is shown in (C) of Fig. 6-9. However, the shoulder of the pipe would create an obstruction if this type of fitting were to be used in a drainage piping. As shown in (B) the DWV copper fitting used with DWV copper tubing or the PVC-DWV fitting used with PVC-DWV pipe provides a smooth passageway.

Durham system is a term used to describe soil or waste systems where all piping is of threaded pipe, tubing, or such other rigid construction, using *recessed drainage fittings* to correspond to the type of piping used.

*Author's note: bottom of drainboard or counter top.

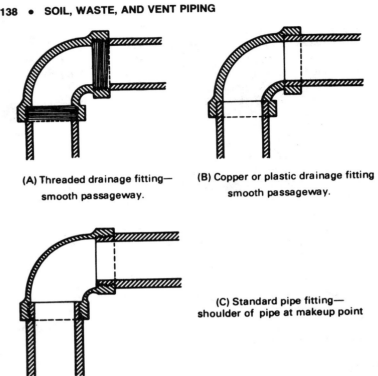

(A) Threaded drainage fitting—
smooth passageway.

(B) Copper or plastic drainage fitting
smooth passageway.

(C) Standard pipe fitting—
shoulder of pipe at makeup point

Fig. 6-9.   Comparison of pressure and drainage fittings.

Prohibited Fittings and Practices

No double hub fitting, single or double tee branch, single or double-tapped tee branch, side inlet quarter bend, running thread, band, or saddle shall be used as a drainage fitting except that a double-hub sanitary tapped tee may be used on a vertical line as a fixture connection. No drainage or vent piping shall be drilled and tapped for the purpose of making connections thereto and no cast-iron soil pipe shall be threaded. No waste connection shall be made to a closet bend or stub of a water closet or similar fixture. The installation of bell trap floor drains and trough urinals, Fig. 6-10, is prohibited.

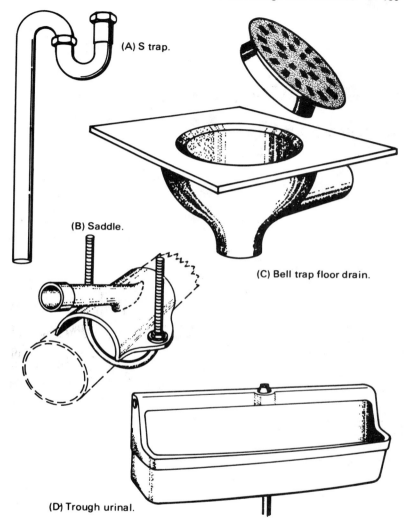

(A) S trap.

(B) Saddle.

(C) Bell trap floor drain.

(D) Trough urinal.

Fig. 6-10. These fittings and fixtures are prohibited.

CHAPTER 7

Working with Cast-Iron Soil Pipe

Cast-iron soil pipe and fittings are manufactured in two grades, service weight and extra heavy. Cast-iron soil pipe is also made in two types, hub and spigot, and plain end pipe and fittings, called No-Hub.

Single-hub soil pipe, Fig. 7-1, is made in 5 ft. and 10 ft. lengths; double-hub soil pipe is made in 2½ ft. and 5 ft. lengths. Hub-and-spigot pipe and fittings may be installed using lead and oakum joints

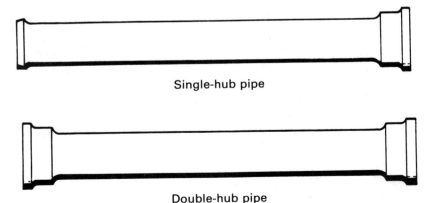

Single-hub pipe

Double-hub pipe

Fig. 7-1. Hub-and-spigot type cast-iron soil pipe.

or, if local codes and job specifications permit, compression type joints, using a neoprene gasket instead of lead and oakum, may be used. Hub-and-spigot cast-iron fittings are made with beaded ends for lead and oakum joints and with plain ends for use in compression joints.

How to Measure Cast-Iron Soil Pipe

There are two basic ways to measure soil pipe, end-to-end and end-to-center measurements. Cast-iron soil pipe is made in two types:

1) Hub-and-spigot pattern.
2) No-Hub pattern.

Fig. 7-2 shows a hub and spigot ¼ bend (90° elbow) inserted

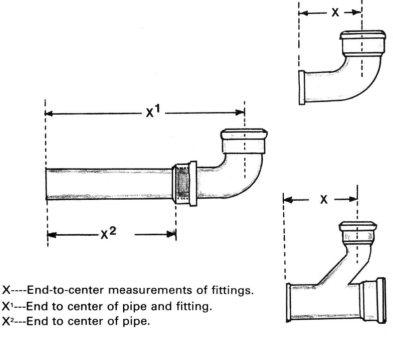

X----End-to-center measurements of fittings.
X^1---End to center of pipe and fitting.
X^2---End to center of pipe.

Fig. 7-2.   How to make end-to-center measurements of soil pipe and fittings.

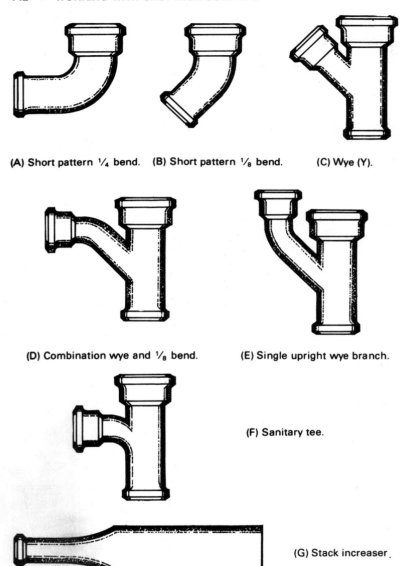

(A) Short pattern ¼ bend.   (B) Short pattern ⅛ bend.   (C) Wye (Y).

(D) Combination wye and ⅛ bend.   (E) Single upright wye branch.

(F) Sanitary tee.

(G) Stack increaser.

Fig. 7-3.   Typical fittings used with lead and oakum soil pipe installation.

into a soil-pipe hub. The point where the spigot end of the ¼ bend rests against the shoulder inside the soil-pipe hub is the make-up point of a soil-pipe hub. The length of the soil pipe is shown as dimension X^2. This is the end-to-end measurement of the pipe and *does not* include the full length of the hub.

In many cases it is more practical to take an end-to-center measurement which includes a fitting. This is shown as dimension X^1 in Fig. 7-2. The end-to-center dimensions of a ¼ bend and a combination wye and ⅛ bend are shown as dimension X.

If the end-to-center measurement of a piece of soil pipe and a ¼ bend needed to connect to a certain point is 35 inches and the takeoff (end-to-center measurement) of a ¼ bend is 8 inches then the end-to-end measurement (X^2) of the cut piece of soil pipe needed would be 27 inches.

NO-HUB soil pipe is also measured in end-to-end and end-to-center measurements but since the pipe and/or fittings butt together there are no hubs to make allowance for.

Cutting Soil Pipe

At some point in the installation of cast-iron soil pipe cut pieces will be needed. Single-hub soil pipe is cast in 5 ft. and 10 ft. lengths; double hub is cast in 2½ and 5 ft. lengths. When a short piece of soil pipe with a hub is needed, it can be cut from a double-hub pipe and the piece of pipe left is not wasted. Soil pipe can be cut using the hammer and chisel method shown in Fig. 7-4; a 12- or 16-oz. hammer and a sharp chisel will be needed. If a hammer and chisel are used, safety glasses should be worn due to the risk of flying cast-iron splinters.

A better and safer method is to use the cutter shown in Fig. 7-7. This cutter has a chain interlaced with cutter wheels and uses pressure applied by the ratchet handle to tighten the chain around the pipe. When the cutter is used, pressure is applied evenly and the resulting cut is smooth with no jagged edges to impede flow through the pipe when the joint is made.

A difficulty encountered in making a joint with a cut piece of double-hub pipe is that there is no bead on the end to center the pipe in the hub. Care must be taken to center the pipe in the mating

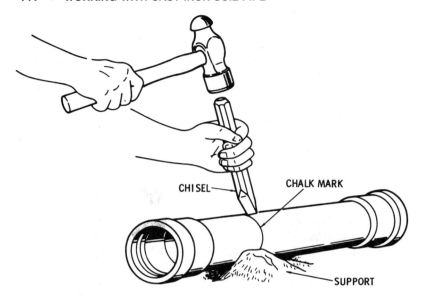

Fig. 7-4.   Method of cutting cast-iron soil pipe using a hammer and chisel.

hub. If the pipe is not centered, it will create an obstruction in the finished joint. Fig. 7-12 illustrates this point.

The tool shown in Fig. 7-7 is designed to cut soil pipe safely. There is no danger from cast-iron splinters as with the hammer and chisel method and the cut will be square and true if the cutter is set properly.

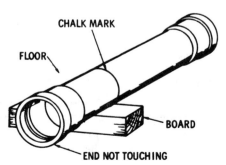

Fig. 7-5.   Cutting the double-hub pipe using a piece of wood to support the pipe.

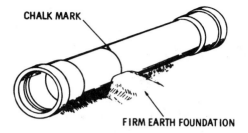

CHALK MARK

FIRM EARTH FOUNDATION

Fig. 7-6.   Cutting double-hub pipe using the ground for support along the cutting line.

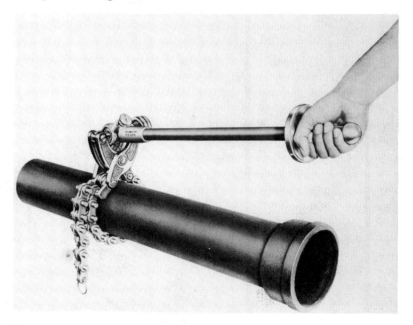

Fig. 7-7.   A ratchet type soil-pipe cutter. *(Courtesy Ridge Tool Co.)*

Lead Joints in Soil Pipe

The first step in making a poured lead joint in soil pipe is to pack the space between the end of the pipe or fitting and the hub with oakum. There are two types of oakum; one is brown in color and is made of hemp fibers which are impregnated with a tarry substance

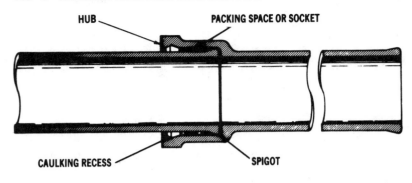

HUB

PACKING SPACE OR SOCKET

CAULKING RECESS

SPIGOT

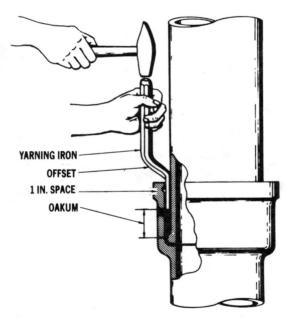

YARNING IRON

OFFSET

1 IN. SPACE

OAKUM

Fig. 7-8.   Packing oakum into a soil-pipe hub.

and twisted; the other is a white oakum made of fiber and impregnated with a powdery material that swells when brought into contact
with water. Water present in the pipe when the pipe is in use will
cause the oakum to swell and prevent leaks. White oakum is pre-
ferred because the joint made with it is virtually leak-proof. The
oakum is rammed into place with a tool called a "yarning" iron, by
hand at first, then packed tightly using a hammer as shown in Fig.
7-8. Special type yarning irons shown in Figs. 7-9 and 7-10 can be
used for making joints in corners or other tight places. A 1-in. space
should be left between the top of the packed oakum and the top of
the hub to receive the molten lead, as shown in Fig. 7-8. The secret
of making a good water-tight joint is to pack the oakum tightly.

For joints near a ceiling, it is necessary to use a ceiling-drop
tool to pack the hub, as shown in Fig. 7-11. This tool is made with
a heavy offset handle which can be struck with a hammer to force
the oakum into the hub. Table 7-1 (p. 148) shows the amount of
oakum and lead needed for various sizes of soil-pipe joints.

When making a horizontal soil-pipe joint a *joint runner*, Fig.
7-13, is used. After the hub has been packed with oakum, the joint

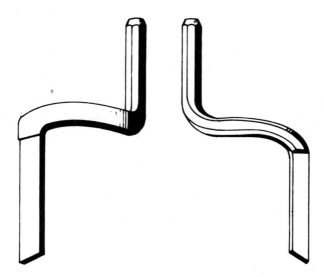

Fig. 7-9.   Left-hand (left) and right-hand (right) yarning irons.

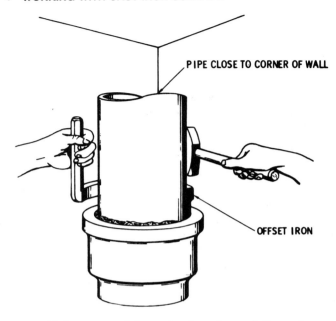

Fig. 7-10. Using a special yarning iron to pack the oakum where the pipe is close to a wall.

runner is placed around the end of the hub and secured with a clamp, leaving a "gate" or small opening to pour lead. The joint runner will hold the lead in the joint until the lead has solidified.

It is sometimes necessary to pour a joint upside down. This may be done by placing the joint runner around the pipe and clamping as shown in Fig. 7-14. A pouring gate is formed by building up a wall of fire clay between the runner and the hub. The lead must be very hot for pouring an upside-down joint.

Table 7-1 Oakum and Lead Requirements for Caulked Joints

Material	Size of Pipe (inches)							
	2	3	4	5	6	7	8	10
Oakum (feet)	3	4½	5	6½	7½	8½	9½	12
Lead (lbs.)	1½	2¼	3	3¾	4½	5¼	6	7½

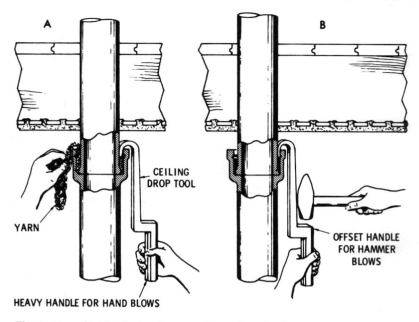

Fig. 7-11.   Method of using a ceiling-drop tool.

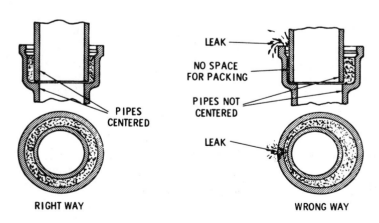

Fig. 7-12.   The right and wrong way of making a joint.

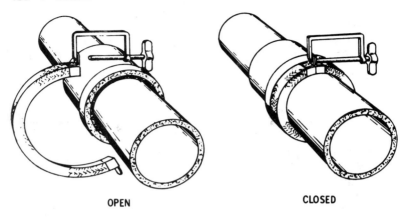

OPEN   CLOSED

Fig. 7-13.   An asbestos joint runner.

Pouring the Lead

Lead is usually melted using a propane tank and burner or a gasoline-fueled firepot. The lead is melted in a cast-iron pot which should be filled to near the top to allow for skimming off the dross present in commercial lead. When lead is at the correct temperature for pouring it will be a bright silver color and will pour easily from the ladle. The ladle used to pour the lead joint should be large enough to hold more lead than the joint will require as the joint should always be filled with one pour. The joint should be dry because any moisture present in the hub could flash into steam on contact with the molten lead, causing a miniature explosion, blowing molten lead out of the joint. A small amount of oil or rosin, poured into a wet hub before pouring the lead, will minimize this risk. When the joint has been poured and the lead has solidified, the joint must be caulked.

Caulking the Lead

After pouring the lead the joint must be caulked or tightened to make the joint water and gas tight. There is more to caulking a joint than packing the lead down into the hub. Two kinds of caulking irons are used; one is an inside iron, Fig. 7-15 (A), the other is an

BELL UPSIDE DOWN

Fig. 7-14. Pouring an upside-down joint with the use of a joint runner.

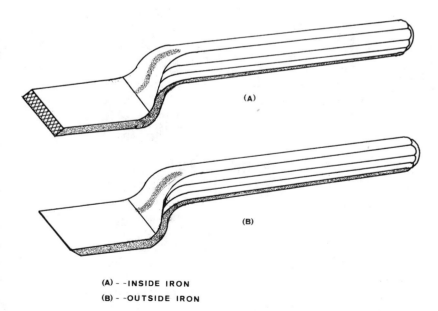

(A)

(B)

(A) - -INSIDE IRON
(B) - -OUTSIDE IRON

Fig. 7-15. Inside and outside caulking irons.

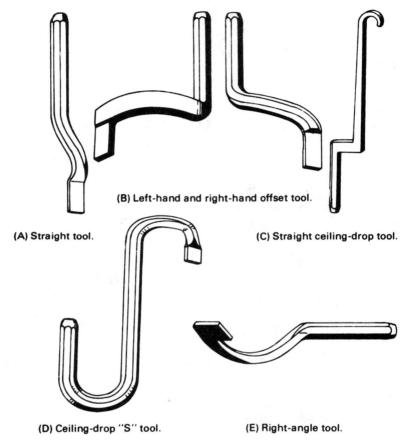

(B) Left-hand and right-hand offset tool.

(A) Straight tool.

(C) Straight ceiling-drop tool.

(D) Ceiling-drop "S" tool.

(E) Right-angle tool.

Fig. 7-16.   Special yarning and caulking tools.

outside iron, Fig. 7-15 (B). The inside iron forces the lead against the pipe on the inside of the hub; the outside iron forces the lead against the inner edge of the hub. A medium-weight (12 oz.) hammer should be used when caulking. If the lead is packed too tightly, the hub can be broken or cracked. Where the joint is fully accessible, regular pattern irons as shown in Fig. 7-15 can be used. In close places, such as in corners or against ceilings, special irons shown in Fig. 7-16 may be needed.

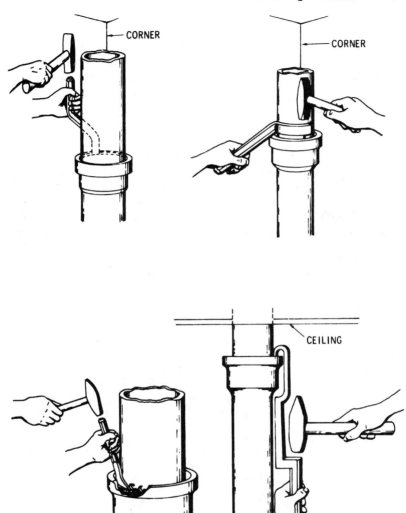

Fig. 7-17. Special caulking tools and their uses.

The Compression Joint

The compression joint is the result of research and development pursued by a number of foundries to provide an efficient, lower-cost method for joining hub-and-spigot cast-iron soil pipe and fittings. The joint is relatively new only in application to cast-iron soil pipe, since similar compression-type gaskets as shown in Fig. 7-18 have been used successfully with cast-iron water main for many years. The compression joint is used with plain (no bead) spigot end cast-iron soil pipe and fittings. Lead and oakum are not used in compression joints; instead, the one-piece rubber gasket shown in Fig. 7-18 is inserted into the soil-pipe hub, a lubricant is applied to the inside of the gasket, and the spigot end of the pipe or fitting is then shoved or pulled, using a tool shown in Fig. 7-19, into the soil-pipe hub. When the spigot end of the pipe is in place in the hub the joint is sealed by displacement and compression of the rubber gasket. The resultant joint is leak proof, root proof, and pressure proof; it absorbs vibration and can be deflected up to five degrees without leakage or failure.

The tool shown in Fig. 7-19 is used to pull pipe and fittings together when making a compression joint. A special type of lubricant can be used to make penetration of the spigot end into the gasketed hub easier.

Fig. 7-18. Type of gasket used to make compression-type soil-pipe joints.

Fig. 7-19. **Soil-type assembly tool used to install compression soil pipe and fittings.** *(Courtesy Ridge Tool Co.)*

The No-Hub Joint

The No-Hub joint for cast-iron soil pipe and fittings is a new plumbing concept which supplements the lead and oakum, and compression-type hub-and-spigot joints by providing another and more compact arrangement without sacrificing the quality and permanence of cast iron. As can be seen in Fig. 7-20, the system uses a one-piece neoprene gasket and a stainless-steel shield and retaining clamps. The advantage of the system is that it permits joints to be made against a ceiling or in any limited-access area. In its 2-in. and 3-in. sizes it will fit into a standard 2″ × 4″ partition without furring.

The stainless-steel shield is noncorrosive and resistant to oxidation, warping, and deformation. It offers rigidity under tension

NO-HUB SOIL PIPE

NEOPRENE GASKET

STAINLESS STEEL CLAMP

NO-HUB SOIL PIPE

Fig. 7-20. A typical connection made between pipe and fittings using stainless-steel clamps with neoprene gaskets. *(Courtesy Cast-Iron Soil Pipe Institute)*

and yet provides sufficient flexibility. The shield is corrugated in order to grip on the gasket sleeve and give maximum compression distribution. The stainless-steel worm gear clamps compress the neoprene gasket to make a permanent, water-tight, gas-tight joint. The gasket absorbs shock vibration and completely eliminates galvanic action between the cast-iron pipe and the stainless-steel shield.* Three typical NO-HUB fittings are shown in Fig. 7-21.

*Reprinted from *Cast-Iron Soil Pipe and Fittings Handbook* by permission.

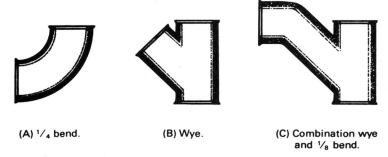

(A) ¼ bend.   (B) Wye.   (C) Combination wye and ⅛ bend.

Fig. 7-21.   No-Hub soil-pipe fittings.

Corrosive-Waste Pipe

The rapid growth of the use of acids and other corrosives in industrial work, as well as the increasing number of schools, colleges, and hospitals containing chemical laboratories, makes it necessary to have a knowledge of special plumbing materials. Among industrial users, the most common are photoengravers, manufacturing jewelers, and industries that manufacture enameled or plated articles and therefore must use acids for cleaning the material.

Several kinds of pipe are used to meet the severe requirements of drainage systems for corrosive wastes, such as:

1. Noncorrosive metal.
2. Plastic.
3. Glass.

Noncorrosive Pipe

An example of noncorrosive pipe is Duriron pipe. This pipe may be cut with a cold chisel and hammer, just like cast-iron soil pipe, but the metal is so hard that the chisel will make only a slight scratch on the pipe. This makes it necessary to go around the pipe two or three times with the chisel and to use about the same weight hammer blows as with cast iron before the pipe will break clean. A pipe cutter having a coil spring above the specially hardened cutter wheel will save much time on a job.

In making joints on Duriron pipe, asbestos rope (at least 85 percent pure) should be used in place of hemp or oakum in order to make an acid-proof joint, and the lead should be poured at as low a temperature as possible. If too hot, the bell or hub may be cracked while caulking.

CHAPTER 8

Roughing-In Plumbing

Roughing-in as applied to plumbing and pipe fitting is a term used for the installation of concealed piping or fittings at the time a building is under construction or being remodeled. As the building nears completion, the final connections of plumbing or heating fixtures are made.

Roughing-in is very important. If the piping is not roughed-in correctly, the final connections may be very difficult. Plans or blueprints show room dimensions and equipment locations, but they do not show at what elevations or locations piping must be stubbed through a wall for final connections.

Roughing-in drawings are furnished by equipment suppliers, and these drawings show not only the correct location of the roughed-in piping but also show where to install any necessary backing boards behind the finished walls. A wall-hung lavatory hangs on a hanger or bracket, and this hanger in turn must be fastened to the wall, using long wood screws that will go through the plaster or drywall and into a backing board.

The plans for a residence may show a bathroom with fixture locations, but usually no definite measurements or locations are given. The location of the stub wall at the end of the bathtub is determined by the length of the tub. This type of bathtub is called a *recessed tub* because it is placed against the stud walls, and the lath and plaster or drywall are then installed.

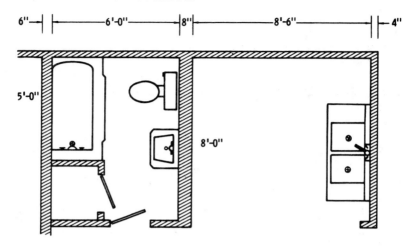

Fig. 8-1. A partial architectural plan of a building showing the kitchen and bathroom areas.

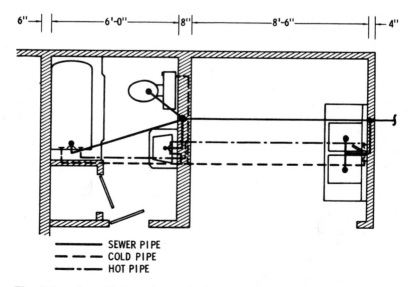

Fig. 8-2. A partial mechanical plan showing the kitchen and bathroom areas.

Fig. 8-1 shows an *architectural* plan of a typical bathroom. Plans and blueprints use the symbol ' to indicate feet and " to indicate inches. The plan in Fig. 8-1 shows the bathroom to be 6'0" wide by 8'0" long; both are finished room measurements. Fig. 8-2 shows a *mechanical* plan of the same bathroom; here again, the fixture locations are shown, water and waste piping is indicated, but no measurements are given. The exact location of the water closet and the lavatory must be determined by the plumber. The minimum space to be allowed for the installation of a water closet is generally conceded to be 32 in. If there is sufficient space, 36 in. is preferable.

As shown in Fig. 8-1 and Fig. 8-2 there is ample room for a 36-in. space for the water closet. At the time the plumber is roughing-in the building, only the stud walls are up, so he must determine where the finished walls will be. If the walls are to be a typical drywall and ceramic tile, for instance, he would add ⅝ in. for drywall and ⅜ in. for ceramic tile and locate the center of the waste piping for the water closet 19 in. from the outside stud wall (18" + 1").

In order to find the distance out from the 8-ft. wall to the center of the same waste piping, the plumber would then check the roughing-in measurements for the water closet. A typical rough-in sheet for a water closet is shown in Fig. 8-3. This shows the center of the waste piping to be 12 in. from the finished wall or 13 in. from the rough wall (12" + ⅝" + ⅜"). Using these methods the center of the waste piping from both walls has now been determined.

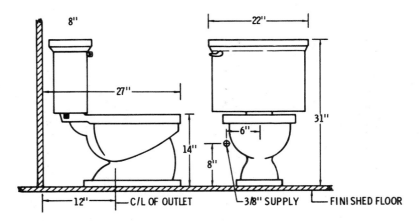

Fig. 8-3. Water closet rough-in sheet.

Rough-In Drawing

Water closet rough-in sheets are similar to Fig. 8-3. All the necessary information for roughing-in a water closet is shown on the roughing-in sheet furnished by the manufacturer. Rough-in sheets vary, depending on the manufacturer and type of fixture, but they will show the important points as outlined in the preceding section.

The center of the lavatory waste piping may be determined in any one of several ways. A medicine cabinet may already be framed in, in which case it would be necessary to center the lavatory on the medicine cabinet for the finished appearance to look right. Another way is to scale the plan. If a rule is laid on the plan, it shows the distance from the outside wall to be 4'-6" to the center of the lavatory. A job condition can sometimes make scaling a plan impractical. In this instance scaling would probably work out very well, due to the size of the room. The center of the lavatory waste piping would be 55 in. from the outside *rough* wall (54" + ⅝" + ⅜").

Fig. 8-4 is a rough-in drawing for a lavatory that is wall-hung. The drawing for this particular make, size, and type of fixture shows the center of the waste piping to be 17 in. from the finished floor. If the plumber is working from a rough floor, he must allow for the thickness of the material which will be added to produce the finished floor. Assuming this to be 2 in. (1⅝" grout + ⅜" ceramic tile), the center of the waste piping would be 19 in. above the *rough* floor.

The center of the waste piping for the bathtub is determined when the waste and overflow fitting is installed on the tub. The drain piping, with a trap, is then extended to this point and connected to the waste and overflow fitting.

Rough-in drawings show the location of the piping needed for final connection to the fixtures. The water supply location for the water closet is shown in Fig. 8-3 as being 6 in. to left of center of the fixture and 8⅞ in. above the finished floor. The mechanical plan of the piping shown in Fig. 8-2 indicates that the water piping to the water closet shall be ½ in. However, the final connection will be ⅜ in. Common practice is to use a reducing fitting (½" × ⅜") behind the finished wall and stub through the wall with ⅜-in. pipe.

The rough-in drawing, Fig. 8-4, shows the water connections to the lavatory to be 8 in. on center (4 in. to each side of center)

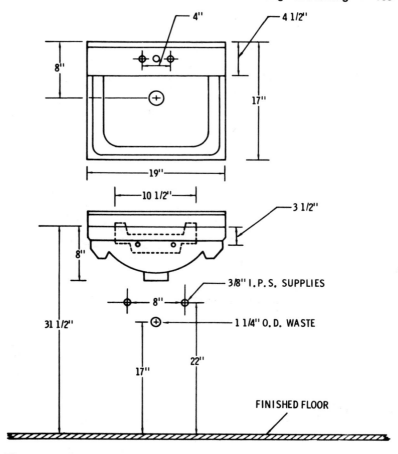

Fig. 8-4. Lavatory rough-in sheet.

and ⅜ in. in size. Here again, it is common practice to use a reducing fitting behind the finished wall and stub through the wall with ⅜-in. pipe.

The rough-in drawing for the bathtub, Fig. 8-5, shows the water piping to be ½ in., 8 in. center to center, and centered on the tub waste and overflow openings. The tub valves, minus the finish chrome-plated parts (trim) and the shower piping, are installed as part of the roughing-in operation. A pipe fitting with a ½-in. f.p.t. (female

BATHTUB ROUGHING-IN SHEETS

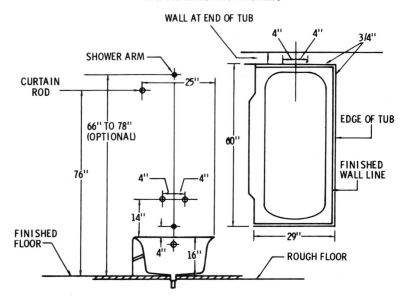

Fig. 8-5.   Bathtub rough-in sheet.

pipe thread) must be left behind the finished wall to receive the chrome-plated shower arm, and the finish trim is installed as part of the finish operation when the fixtures are set.

A typical roughing-in sheet for a bathtub shows that backing boards are usually necessary for securing the curtain rod to the wall, as well as the piping rough-in locations. The tub faucet, or mixing valve (called an over-rim filler), should be centered on the waste and overflow connection.

Recessed tubs are identified as L.H. (left-hand) or R.H. (right-hand). Facing the tub front, or apron, if the drain connection is at the left end of the tub, it is a L.H. tub. If the drain is at the right end, it is a R.H. tub, as shown in Fig. 8-6. Corner tubs are identified in a different manner. Facing the front or apron of the tub, if the corner is on the left end, it is a L.H. corner tub; if the corner is on the right, it is a R.H. corner tub.

A typical kitchen sink rough-in sheet does not show water or waste opening rough-in dimensions. Job conditions plus a great variety in the selections of faucets and drain connections make it

Fig. 8-6. **A right-hand recessed bathtub.** *(Courtesy Eljer Plumbing-ware)*

impractical to show the rough-in measurements. In almost all cases, the suggested measurements shown in Fig. 8-7 will enable connection to either a double sink waste or to a single waste and a garbage disposal. If the waste opening is roughed in at approximately the center of either of the sink compartments, final connection will be easier. The hot- and cold-water piping can be roughed-in through the wall at approximately 16 in. above the finished floor or roughed-in through the floor for final connection.

Isometric Drawing

The purpose of an isometric drawing is to give a view of all parts of a system. If you can imagine that you are standing above and slightly to one side of the bathroom we have been describing and that you are looking down on the finished roughed-in piping, it would look similar to the view shown in Fig. 8-8. Isometric drawings are a

KITCHEN SINK ROUGH-IN SHEET

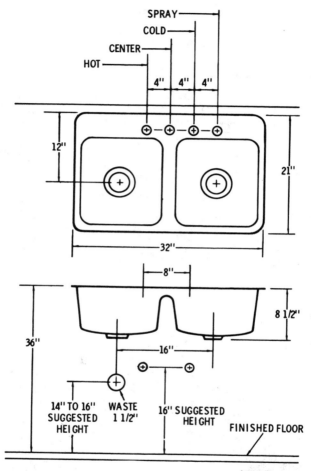

Fig. 8-7.   Kitchen sink rough-in sheet.

valuable aid to estimating or ordering fittings since virtually every fitting used on the job will be shown in the drawing, if the drawing is done correctly. Fig. 8-8 is an isometric drawing showing which part of the soil, waste, and vent piping uses cast-iron soil pipe and which part uses DWV (drainage, waste, and vent) copper piping.

Roughing-in a building is not a big job—it is a series of little

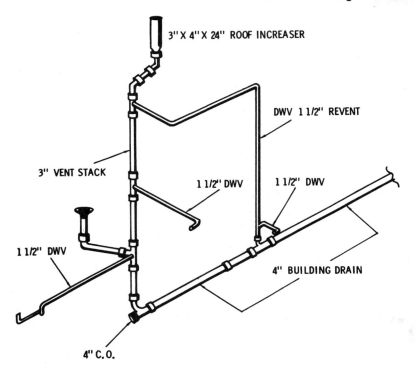

3" X 4" X 24" ROOF INCREASER

DWV 1 1/2" REVENT

3" VENT STACK

1 1/2" DWV

1 1/2" DWV

1 1/2" DWV

4" BUILDING DRAIN

1 1/2" DWV

4" C.O.

Fig. 8-8.   A typical isometric drawing.

jobs. A ten-story building seems like a big job, but if one looks at it as ten one-story jobs, it doesn't seem so formidable. Fig. 8-2 shows a partial mechanical plan of a bathroom and kitchen area of a residence. The experienced plumber can take the plan in Fig. 8-2, make a sketch similar to the one in Fig. 8-9, and from this sketch order all the soil, waste, and vent material needed to do the job.

Fig. 8-9 is an isometric drawing showing hub-and-spigot type cast-iron soil pipe and DWV weight copper tubing. Hub-and-spigot cast-iron soil pipe is made in single-hub and double-hub patterns. Single hub is available in both 5 ft. and 10 ft. lengths; double hub is available in 2½ ft. (30 in.) and 5 ft. lengths. The 30-in. lengths provide short pieces with hubs, with a minimum of waste. Fig. 8-9 also shows that the building drain and the 3-in. waste and vent stack are cast-iron soil pipe; the 2-in. and smaller waste and vent piping

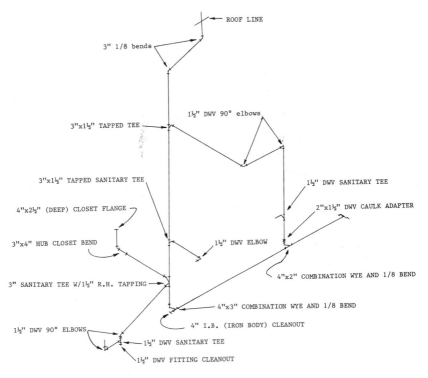

Fig. 8-9. An isometric drawing showing hub-and-spigot cast-iron soil pipe and DWV copper tubing.

is DWV weight copper tubing. The water piping is made up using Type L copper tubing and wrought copper fittings. Most city and state plumbing codes require that the horizontal building drain be 4-in. The vertical soil, waste, and vent stack which receives the discharge of the lavatory, bathtub, and water closet (toilet) can be sized at 3 in. All piping is sized as inside diameter.

Material Lists

Using the information given in Fig. 8-9 a material list for the job can be made.

Fixtures:

1—5 ft. L.H. (left-hand) recess tub with over-rim filler (faucets), shower head, and trip lever drain.

1—12″ rough-in reverse trap closet combination with white seat, flexible supply, and C.P. stop.

1—19″ × 17″ V.C. (vitreous china) lavatory with 4″ centerset faucet, pop-up drain, supplies, stops, and C.P. P trap.

1—32″ × 21″ stainless-steel sink, complete with basket strainers, connected sink waste, trap, and single-lever faucet with spray.

Material:

10 ft. (1 length) 4 in. S.H. (single-hub) soil pipe.

5 ft. 4 in. D.H. (double-hub) soil pipe.

5 ft. 3 in. S.H. soil pipe.

10 ft. 3 in. D.H. soil pipe (5′ lengths).

2—30-in. lengths 3-in. D.H. soil pipe.

(It is always better to order a little extra soil pipe to allow for cracked hubs, etc.; the extra material can always be returned to the shop.)

1—4″ × 2″ combination wye and ⅛ bend.

1—4″ × 3″ combination wye and ⅛ bend.

1—3 in. sanitary tee with 1½-in. R.H. tapping.

1—3″ × 4″ hub closet bend.

1—4″ × 2½″ (deep) closet flange.

2—3″ × 1½″ tapped sanitary tees.

2—3 in. ⅛ bends.

1—3″ × 4″ × 24″ calk roof increaser.

1—4 in. I.B. (iron-body) cleanout.

The amount of lead is estimated in this manner: 1 lb. for each inch/joint. A 3-in. joint would need 3 lbs. of lead. Oakum is estimated at 1 lb. of oakum for each 10 lbs. of lead. In actual practice it will be found that the amount of lead used will be somewhat less than the 1 lb. per inch formula but this formula will allow for spillage, etc. In the partial plan we are working with we would have, according to our isometric drawing (Fig. 8-9):

7—4-in. lead joints = (7 × 4)   28 lbs. lead
12—3-in. lead joints = (12 × 3) 36 lbs. lead
1—2-in. lead joint = (1 × 2)   2 lbs. lead
                        Total 66 lbs. lead

The amount of oakum needed, using the formula given above, would be 7 lbs. (rounding off the 66 lbs. of lead to the nearest tenth). The waste and vent piping for the kitchen sink would require:

2—1½ in. DWV 90° elbows.
1—1½ in. DWV M.I.P. (male iron pipe) copper adapter.
1—1½ in. DWV copper sanitary tee.
20 ft. (approx.)—1½ in. DWV copper pipe.

The waste piping for the lavatory would require:

1—1½ in. DWV M.I.P. copper adapter.
1—1½ in. DWV 90° elbows.
3 ft. (approx.)—1½ in. DWV copper pipe.

The waste piping for the bathtub would require:

1—1½ in. DWV M.I.P. adapter.
2—1½ in. DWV copper 90° elbows.
1—1½ in. DWV copper tee.
1—1½ in. DWV copper fitting cleanout.
6 ft. (approx.)—1½ in. DWV copper pipe.

Approximately ½ lb. of 50/50 solder would be required for the DWV waste and vent piping. If the waste and vent piping were installed as shown in the drawings, the bathtub and water closet would be wet-vented by the lavatory, which would be permissible by virtually all local plumbing codes. A certain amount of knowledge will be called for when estimating the water piping material needed. For instance, valves may not show on the plans at the bathtub and kitchen sink locations, but good plumbing practice (and most state and city codes) demands that valves be installed at these locations. Final connections at the bathtub and kitchen sink locations will require ½-in. openings; the water closet and the lavatory will require ⅜-in. openings.

Water piping material for the four fixtures shown:

2—¾″ × ¾″ × ½″ copper tees.
1—¾″ × ½″ × ¾″ copper tee.

2—¾" × ½" × ½" copper tees.

1—¾" copper 45° elbow.

1—½" copper 45° elbow.

6—½" copper 90° elbows.

3—½" copper × ⅜" I.P.S. iron pipe size elbows.

2—½" copper sweat stops.

30 ft. (±) ½ in. hard copper tubing. (type K, L, or M).

15 ft. (±) ¾ in. hard copper tubing. (type K, L, or M).

½ lb. 95/5 soft solder.

Sand cloth and solder flux; or instead of soft solder, Silfos or similar "silver solder."

Recent studies have shown that lead from lead/tin solders will leach into water passing through lead-soldered joints, due to solder

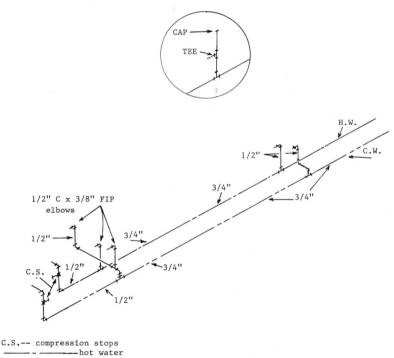

C.S.— compression stops
——— - ———hot water
———--———-- — cold water

Fig. 8-10.   An isometric drawing of the hot- and cold-water piping shown in Fig. 8-2.

flowing into the inside of pipe and fittings when joints are made. The possibility of lead poisoning from this source can be avoided by using one of the so-called silver solders instead of the lead/tin solders.

Fig. 8-10 is an isometric drawing of the water piping shown in Fig. 8-2. The inset shows an air chamber installed on a fixture branch. Air chambers are designed to prevent water hammer, the noise created when a faucet is closed suddenly or when a solenoid valve on a clothes washer or dishwasher closes.

Water is forced through piping by pressure impressed on the water main. When this flow is stopped *suddenly*, a shock wave is created, the wave races back in the piping, creating for an instant,

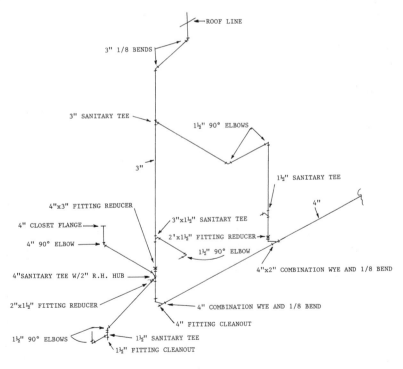

Fig. 8-11.   An isometric drawing showing PVC-DWV piping and fittings.

a higher-than-incoming pressure. Air which is trapped in the top of an air chamber will compress and absorb the shock wave and thus eliminate the water hammer.

An air chamber can be constructed by using a tee instead of an elbow at the top of a fixture branch and extending the pipe upwards 12 to 18 inches, with a cap at the top. Air chambers should be installed on both hot and cold water piping. Air chambers or water hammer arrestors are required by some building codes. Commercial models are available through plumbing supply houses.

PVC-DWV Waste and Vent Piping

The following material would be needed to install the PVC-DWV waste and vent piping shown in Fig. 8-11:

1—4″ PVC-DWV combination wye and ⅛ bend.
1—4″ × 2″ PVC-DWV combination wye and ⅛ bend.
1—4″ PVC-DWV fitting cleanout.
1—4″ PVC-DWV sanitary tee with 2″ R.H. hub.
1—4″ × 3″ PVC-DWV fitting reducer.
1—4″ PVC-DWV 90° elbow.
1—4″ PVC-DWV closet flange.
2—3″ × 1½″ PVC-DWV sanitary tees.
2—3″ PVC-DWV ⅛ bends (45° elbows).
2—2″ × 1½″ PVC-DWV fitting reducers.
1—1½″ PVC-DWV fitting cleanout.
2—1½″ PVC-DWV sanitary tees.
5—1½″ PVC-DWV 90° elbows.
20′—4″ PVC-DWV (schedule 40) pipe.
30′—1½″ PVC-DWV (schedule 40) pipe.
1 pint PVC cleaner.
1 pint PVC cement.

Fig. 8-11 shows the material needed to rough-in waste and vent piping for the areas shown in Fig. 8-2 if PVC-DWV piping and fittings are used. Schedule 40 PVC-DWV piping and fittings are widely accepted and used as an alternate to cast iron, copper tubing, and steel pipe as waste and vent piping.

The importance of accurate measurements for roughing-in purposes was explained in Chapter 5, Blueprints and Elevations. It is

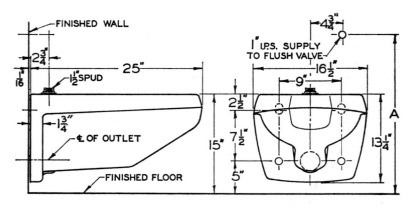

Fig. 8-12. Roughing-in measurements for a wall-hung closet.

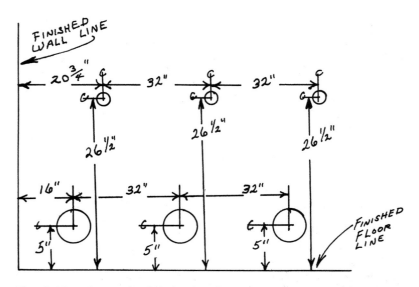

Fig. 8-13. A rough sketch showing water and waste openings, used when roughing-in toilet room.

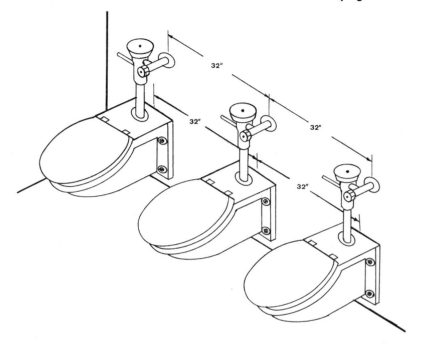

Fig. 8-14. Fixtures will be perfectly aligned if rough-in is correct.

not practical to carry a large set of prints when roughing-in a job; they are easily torn and will fade if exposed to sunlight. It is far better to take the time to make a rough sketch showing all the important measurements needed to rough-in the piping. Referring to Chapter 5, Fig. 5-3 shows three wall-hung water closets. The water closets will be enclosed by toilet partitions 32 in. on center; therefore, the water closets must be set 32 in. on center also. The rough sketch, Fig. 8-13, shows all the information needed to rough-in the water and waste connections for these fixtures.

Special soil-pipe fittings are used for the waste connections and the tappings can be ordered in various positions as shown in Fig. 8-15. As the waste line slopes upward at a pitch of ¼ inch per foot, the tappings are progressively lower. The fittings shown in Fig. 8-15 are made in R.H. (right-hand) and L.H. (left-hand) patterns, and are identified by placing the fitting as shown in Fig. 8-16 with

Tappings and Dimensions

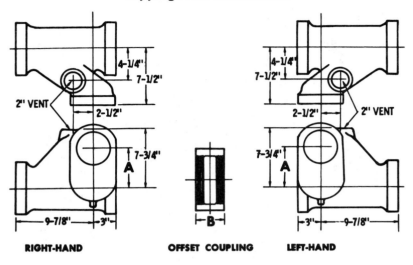

RIGHT-HAND   OFFSET COUPLING   LEFT-HAND

4″ × 4″ and 5″ × 4″	
Tapping No.	Dimension A
1	$5\frac{1}{4}$
2	$4\frac{7}{8}$
3	$4\frac{1}{2}$
4	$4\frac{1}{8}$
5	$3\frac{3}{4}$
6	$3\frac{3}{8}$
7	3
8	$2\frac{5}{8}$
9	$2\frac{1}{4}$
10	$1\frac{7}{8}$
11	$1\frac{1}{2}$
12	$1\frac{1}{8}$

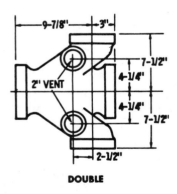

DOUBLE

Fig. 8-15.   Positions of waste openings in fittings.

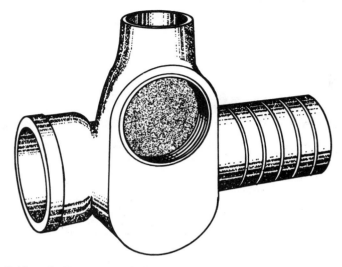

Fig. 8-16. A 4″ tapping is shown in position 1.

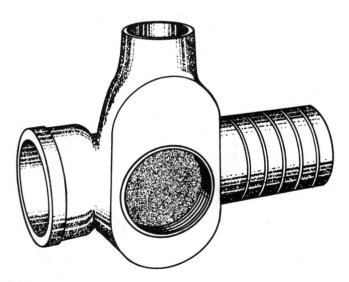

Fig. 8-17. A 4″ tapping is shown in position 5.

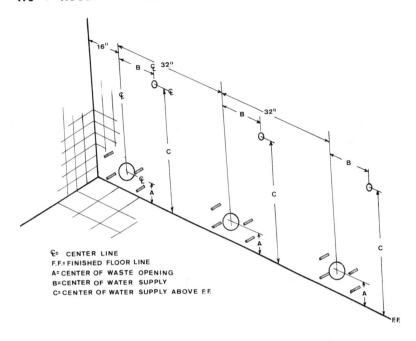

Fig. 8-18.   Finished wall ready to receive closet bowls.

₵= CENTER LINE
F.F.= FINISHED FLOOR LINE
A= CENTER OF WASTE OPENING
B= CENTER OF WATER SUPPLY
C= CENTER OF WATER SUPPLY ABOVE F.F.

the waste opening facing the viewer. If the spigot end is at the right it is a R.H. flow fitting; if the spigot is at the left it is a L.H. flow fitting. When this type of fitting is used, the waste piping will be placed above floor levels, lying on the floor in the pipe space behind the toilet room wall. A device called a chair carrier, shown in Chapter 9, Fig. 9-15, is used to mount the toilet bowl to the wall. Four ⅝-in. bolts project from the concealed barrier through the finished wall as shown in Fig. 8-18. If the fixture carriers are installed correctly, the closet bowls will be mounted in perfect alignment and spacing as shown in Fig. 8-14. A typical rough-in sheet for a water closet is shown in Fig. 8-12.

CHAPTER 9

Plumbing Fixtures

When a new building or a remodeling project is near completion, the plumbing fixtures must be installed and connected. These fixtures usually consist of, but are not limited to, the following:

Lavatories.
Water closets.
Bathtubs, with or without showers.
Sinks: Kitchen, Laundry, Laboratory.
Urinals.
Bidets, common in Europe, coming into use in the United States.

Each kind of fixture and the materials of which they are made are shown below.

Vitreous China

Vitreous china fixtures are made of pottery clays, formed in molds and allowed to air dry. After drying, the fixture is sprayed with a silicone coating, then placed in an oven and baked at extremely high temperatures. This transforms the silicone into porcelain that has a glassy, glossy finish. Porcelain is a form of glass and must be treated accordingly; water closet bowls must be tightened down evenly when being set since there is no "give" in vitreous china. This also applies to wall-hung vitreous china urinals. The porcelain finish of vitreous china fixtures is very hard and scratch-resistant and will give many years of service if handled properly.

Fixtures made of vitreous china:

Lavatories.
Water closets.
Sinks.
Bidets.
Urinals.

Lavatories

Lavatories are available in many different styles and are made of vitreous china, porcelain-enameled pressed steel, and porcelain-enameled cast iron. The most common types are shown in Figs. 9-1 and 9-2. Fig. 9-1 shows a vitreous china counter-top model with an 8-in. centerset faucet and pop-up drain. Counter-top models are

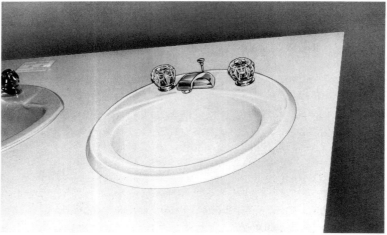

Fig. 9-1. A self-rimming vitreous china lavatory. *(Courtesy Crane Plumbing)*

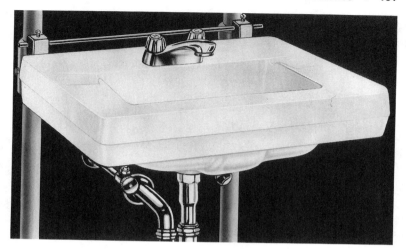

Fig. 9-2. **A wall-hung vitreous china lavatory.** *(Courtesy Crane Plumbing)*

made in both white and colors and are used principally in residential bathrooms.

The lavatory shown in Fig. 9-2 is a vitreous china wall-hung model, used principally in public buildings, schools, and office toilet rooms. The unit can be hung on a wall-mounted bracket or, for greater strength, secured to a concealed carrier arm which is attached to a steel plate behind the wall. The model shown in Fig. 9-2 has a 4-in. centerset faucet and a pop-up drain. Lavatories installed in public buildings are usually equipped with open strainer drains instead of pop-up drains.

P traps (1½ in. or 1¼ in.) are used to connect the drain piping, and the hot and cold water connections can be made using flexible type supplies as shown in Fig. 9-3 and Fig. 9-4. The construction of a typical pop-up lavatory drain is shown in Fig. 9-4.

The lavatory shown in Fig. 9-5 is a hospital/clinic fixture with foot pedal faucet controls. The foot pedal controls diminish the risk of contamination after washing hands. Porcelain-enameled pressed steel lavatories are relatively inexpensive and chip easily. Porcelain-enameled cast-iron lavatories are very durable and are comparable in price to vitreous china lavatories.

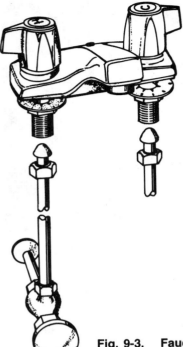

Fig. 9-3. Faucet connections made with flexible supply tubes.

Water Closets

There are two basic types of water closets:

1. Floor mounted; bottom outlet, back outlet.
2. Wall-hung; flush valve (flushometer type), tank type.

There are three basic types of water closet bowls:

1. Siphon jet.
2. Reverse trap.
3. Blow-out.

Closet bowls are made in two basic designs:

1. Round front bowls.
2. Elongated bowls.

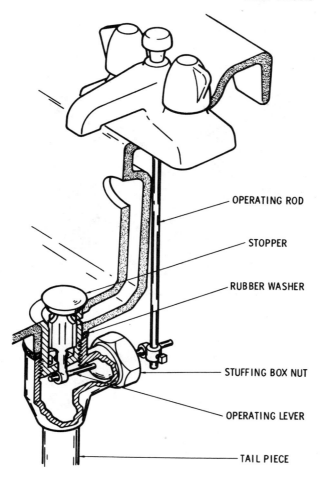

OPERATING ROD

STOPPER

RUBBER WASHER

STUFFING BOX NUT

OPERATING LEVER

TAIL PIECE

Fig. 9-4.   A typical pop-up lavatory drain.

Ecological concerns have brought about the development of water saver type closet combinations. (Closet combinations is the term used when a closet tank is mounted on a closet bowl). A floor mounted closet combination with a bottom outlet and a water saver tank is shown in Fig. 9-6. This type closet combination is used primarily in residences and office toilet rooms.

Fig. 9-5. A hospital type vitreous china lavatory with foot controls. *(Courtesy Crane Plumbing)*

Fig. 9-6. A floor mount closet combination with water saver tank. *(Courtesy Eljer Plumbingware)*

Fig. 9-7.   A flushometer/tank type water closet. *(Courtesy Crane Plumbing)*

A water saver type closet combination which is a radical departure from the conventional type is shown in Fig. 9-7. This closet combination derives its flushing power from supply line pressure. This is accomplished through the combination of supply line pressure and a sealed "flushometer tank" located inside of the conventional-appearing tank. The purpose of this design is water conservation and this design enables the closet to flush using only six qts. (1½ gals.) per flush. This closet combination is made with a built-in vacuum breaker and meets all existing local, regional, and national performance and safety codes.

One-piece water closets are designed for near-silent flushing and filling action. A one-piece water closet is shown in Fig. 9-8.

The flushing action in the siphon-jet bowl is accomplished by directing a jet of water through the upward leg of the trapway, which fills the trapway and starts the siphoning action. The strong,

Fig. 9-8.   A one-piece water closet. *(Courtesy Crane Plumbing)*

quick, and relatively silent action of the siphon-jet bowl, together with its deep water seal and large water surface, is recognized by sanitation authorities as the most satisfactory closet bowl in existence.

Blow-out—Since the blow-out type of toilet bowl (Fig. 9-10) depends entirely on a driving jet action for its efficiency, rather than on siphoning action in the trapway, it cannot be compared favorably with other types of toilet bowls. It is economical in its use of water; however, it does have a large water surface that reduces fouling space, a deep water seal, and a large unrestricted trapway. The blow-out bowls are well suited for use in schools, offices, and public buildings. They are operated only by flush valves.

The reverse trap closet bowl, Fig. 9-11, is similar in appearance and flushing action to the siphon-jet bowl. The water surface and size of the trapway are smaller and less water is required for flushing action.

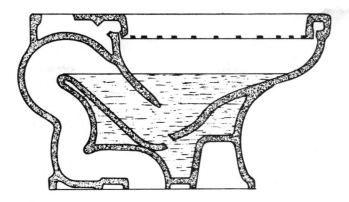

Fig. 9-9. Siphon-jet closet bowl.

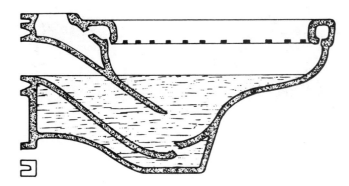

Fig. 9-10. A blow-out closet bowl.

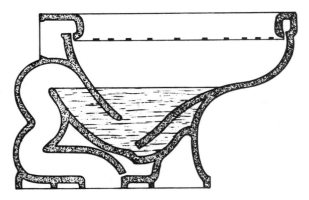

Fig. 9-11. A reverse trap closet bowl.

Fig. 9-12.   A floor mount back outlet closet combination. *(Courtesy Eljer Plumbingware)*

The water closet shown in Fig. 9-12 is designed for use where under-floor space is a problem for structures with little or no crawl space; for structures built on concrete slabs; or in remodeling projects. This fixture has a back outlet, centered 4 in. (±) above finished floor line, which allows the soil pipe to which the fixture is connected to be installed horizontally, above the finished floor. The fixture is bolted to a standard closet flange extending through the finished wall.

A wall-hung tank type closet combination is shown in Fig. 9-13. This fixture has a back outlet and is secured to a fitting (chair carrier, Fig. 9-15) mounted behind the wall. This fixture is used primarily in office and semi-public toilet rooms as the fixture design makes floor cleaning easy.

Wall-hung flushometer type, elongated closet bowls, shown in Fig. 9-14, are chosen by architects and engineers for installation in public buildings, schools, and factories. This type closet bowl is secured to a carrier fitting, Fig. 9-15, mounted behind the wall.

Fig. 9-13. A wall-hung tank type closet combination. *(Courtesy Eljer Plumbingware)*

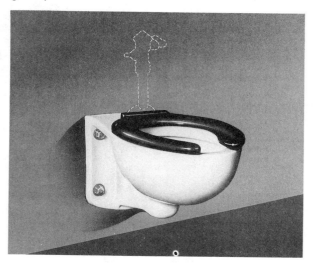

Fig. 9-14. A wall-hung flushometer type closet bowl. *(Courtesy Crane Plumbing)*

Concealed supporting-chair unit for:

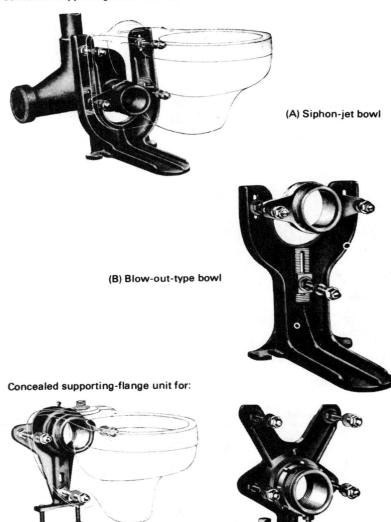

(A) Siphon-jet bowl

(B) Blow-out-type bowl

Concealed supporting-flange unit for:

(C) Blow-out-type bowl

(D) Siphon-jet bowl

Fig. 9-15. Concealed chair carriers for closet bowls. *(Courtesy Crane Plumbing)*

SLOAN ROYAL (DIAPHRAGM TYPE) FLUSH VALVE PRIOR TO MID-YEAR 1971

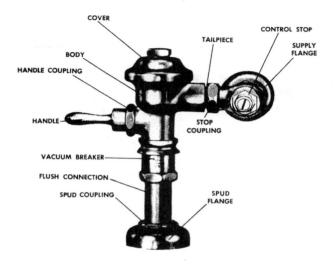

SLOAN ROYAL (DIAPHRAGM TYPE) FLUSH VALVE SINCE MID-YEAR 1971

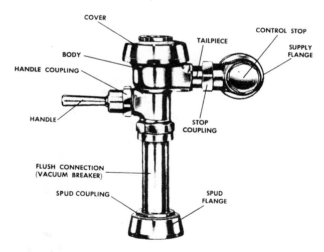

Fig. 9-16.   Diaphragm type flush valves. *(Courtesy Sloan Valve Co.)*

Flushometer Valves

One widely used type of flushometer valve is shown in Fig. 9-16. One model is used on water closets, another for flushing urinals. This diaphragm type requires no regulation to maintain flushing accuracy. When the valve is in closed position the segment diaphragm divides the valve into an upper and lower chamber with equal pressure on both sides of the diaphragm. The greater pressure area on top of the diaphragm holds it closed on its seat. Movement of the flushing handle in any direction pushes the plunger, which tilts the relief valve and allows water to escape from the upper chamber into the lower chamber. Water pressure raises the working parts, the relief valve, disc, diaphragm, and guide, admitting water

VALVES AND FAUCETS

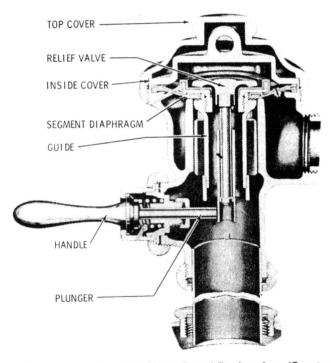

TOP COVER

RELIEF VALVE

INSIDE COVER

SEGMENT DIAPHRAGM

GUIDE

HANDLE

PLUNGER

Fig. 9-17.   A cut-away view of a Sloan Royal flush valve. *(Courtesy Sloan Valve Co.)*

under full pressure and flushing the fixture. While the fixture is flushing, a small hole in the diaphragm permits water pressure to build on the top side of the diaphragm. When the pressure is equalized in both sides of the chamber, the diaphragm is returned to closed position. When this type of valve will not shut off after flushing, it is probably because the equalizing hole in the diaphragm is clogged. The UNIFORM PLUMBING CODE states, as do most plumbing codes, that: *Water closet flushometer valves shall be equipped with an approved vacuum breaker. Each such device shall be installed on the discharge side of the flushometer valve with the critical level at least six (6) inches above the overflow rim of the bowl.* It is also worth noting that the UNIFORM PLUMBING CODE, as do most codes, states: *Water closets for public use shall be elongated bowls equipped with open front seats.*

A cut-away view of a Sloan flush valve is shown in Fig. 9-17. Water connections to tank type water closets are usually offset. The flexible tank connector shown in Fig. 9-18 simplifies this connection.

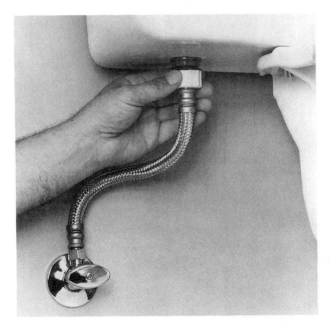

Fig. 9-18.   A flexible connection can be used on water closet tank.
(Courtesy Fluidmaster, Inc.)

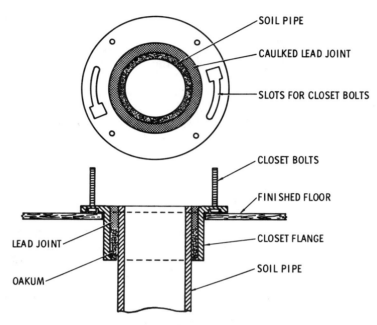

Fig. 9-19. Cast-iron flange connected to cast-iron soil pipe using a lead and oakum joint.

The different types of closet flanges and their correct installation is shown in Figs. 9-19, 9-20, 9-21, 9-22, and 9-23.

Bathtubs

Bathtubs are made in various sizes and colors and of porcelain-enameled pressed steel, porcelain-enameled cast iron, and fiberglass. The most used size is the 5 ft. recessed (set into the wall) tub, available either with R.H. (right-hand) or L.H. (left-hand) drain. When facing the apron or front of the tub and the drain and overflow opening is on the left, it is a L.H. tub. If the drain and overflow opening is on the right, it is a R.H. tub. As to material there are advantages and disadvantages with each type.

Porcelain-enameled pressed steel tubs are the least expensive, but the enameled surface can be chipped easily if a heavy or sharp object falls or is dropped into the tub.

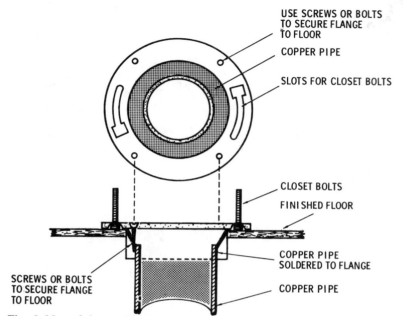

Fig. 9-20.   A brass flange soldered to a copper pipe.

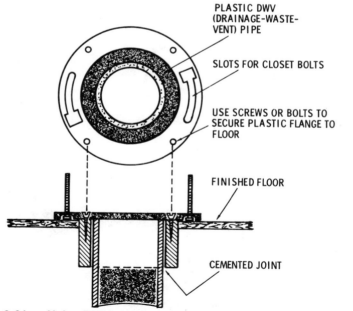

Fig. 9-21.   Using DWV plastic pipe and a DWV plastic closet/flange.

195

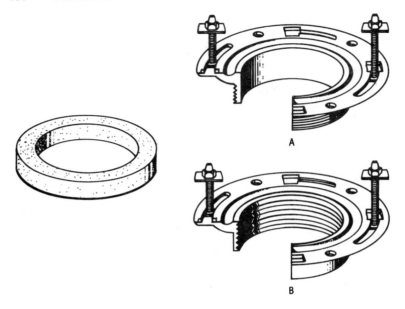

Fig. 9-22.   Threaded brass flanges; (A) male, (B) female.

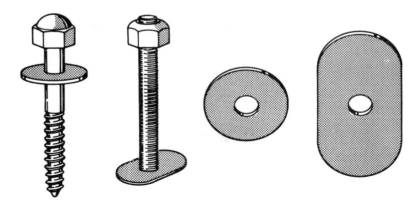

Fig. 9-23.   Closet bolts, screws, and washers.

Fig. 9-24. **A 5′ L.H. recessed bathtub.** *(Courtesy Eljer Plumbingware)*

A cast-iron enameled tub is more durable, less easily chipped, but costs approximately twice as much as a steel tub. In outward appearance both pressed steel and cast-iron tubs are similar to the fixture shown in Fig. 9-24.

Fiberglass tubs and shower enclosures are lightweight, comparable in price to cast-iron tubs, and, as shown in Fig. 9-25, are an attractive addition to a bathroom.

The three-valve faucet shown in Fig. 9-26 is called a transfer valve, directing water into either the tub or the shower by turning the center valve. The two-valve filler is used only to fill the tub. The two-valve shower fitting can be installed above a two-valve filler

Fig. 9-25. A fiberglass bathtub and shower enclosure. *(Courtesy Owens-Corning Fiberglass)*

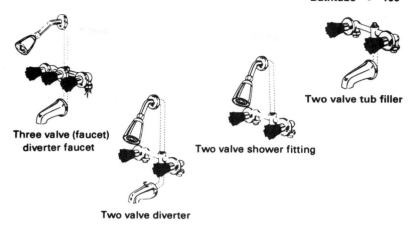

Three valve (faucet)
diverter faucet

Two valve diverter

Two valve shower fitting

Two valve tub filler

Fig. 9-26.   Various types of tub and shower valves.

Fig. 9-27.   A single-handle tub and shower valve.
(Courtesy   Moen/Stanadyne Group)

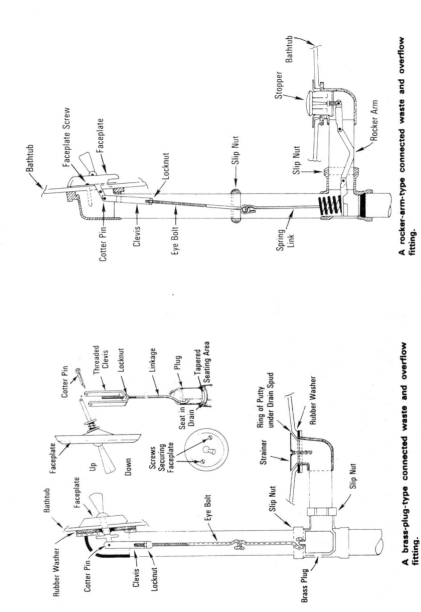

A brass-plug-type connected waste and overflow fitting.

A rocker-arm-type connected waste and overflow fitting.

Fig. 9-28. The most common types of waste and overflow fittings.

to control water to the shower only. (Valves controlling water to bathtubs are called over-rim fillers.) The single handle mixing valve can be used as a tub filler, or by lifting the knob on the spout, can be used to direct water to the shower head. A typical single-handle tub and shower valve is shown in Fig. 9-27. Nearly all of the waste and overflow fittings used on bathtubs are one of the types shown in Fig. 9-28.

Kitchen Sinks

Kitchen sinks are made of porcelain-enameled cast-iron, stainless (nickel) steel, and porcelain-enameled pressed steel.

Cast Iron

Porcelain-enameled cast-iron sinks are durable and chip resistant but the finish can be damaged by abrasive cleaners commonly used in kitchens. Cost is approximately the same as a medium-grade stainless-steel sink. A cast-iron enameled sink is shown in Fig. 9-29.

Fig. 9-29.   A two-compartment porcelain-enameled cast-iron sink.

Fig. 9-30. A two-compartment stainless-steel sink. *(Courtesy Moen/ Stanadyne Group)*

Stainless Steel

Stainless-steel sinks are the most popular type and can be easily cleaned. Price depends on the finish or polish and the gauge or thickness of the metal. As a general rule, the thicker the gauge (number of sheets of metal per inch) and the higher the polish, the more expensive the sink. A typical high-polish stainless-steel sink with single-lever faucet is shown in Fig. 9-30.

Enameled Steel

Porcelain-enameled steel sinks are the least expensive type and can chip easily if heavy or sharp objects, common in the kitchen, are dropped into them. The finish can also be damaged by abrasive cleaners.

Cartridge type single lever faucet.

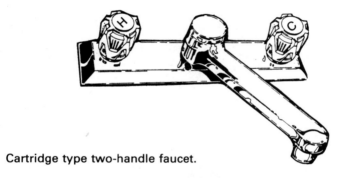

Cartridge type two-handle faucet.

Fig. 9-31. Two types of cartridge type washerless sink faucets.
(Courtesy Moen/Stanadyne Group)

Kitchen Sink Faucets

The faucets shown in Figs. 9-31 and 9-32 are "washerless" faucets. Washerless faucets are made with replaceable seats or replaceable cartridges to control the flow of water. The advantages of the washerless faucet over the bibb washer type are: ease of operation, longer life, and fewer and less expensive repairs.

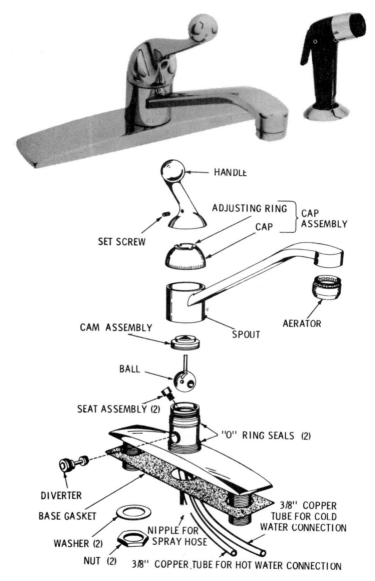

Fig. 9-32. This type of single-lever faucet uses replaceable ball and seats. *(Courtesy Delta Faucet Co.)*

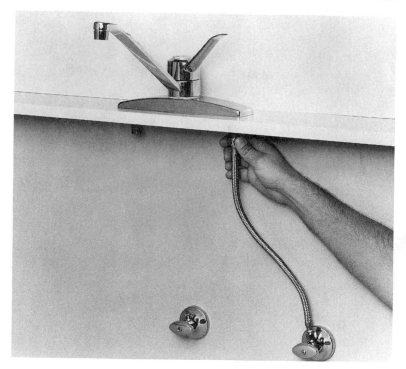

Fig. 9-33. **Flexible connections for a sink or lavatory faucet.** *(Courtesy Fluidmaster, Inc.)*

Flexible Supplies

Flexible supply tubes, Fig. 9-33, can be used to make hot and cold water connections to many sink faucets. Faucets which are made with ⅜ in. O.D. tubing supplies can be connected with the couplings shown in Fig. 9-34.

Urinals

The three principal types of urinals are:

1. Stall urinals.
2. Wall-hung urinals.
3. Pedestal urinals.

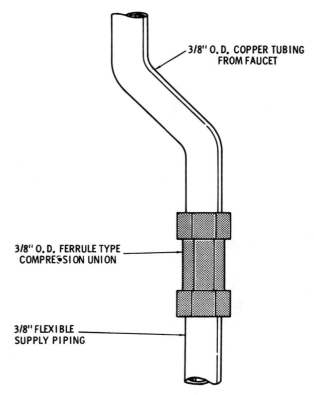

3/8" O. D. COPPER TUBING
FROM FAUCET

3/8" O. D. FERRULE TYPE
COMPRESSION UNION

3/8" FLEXIBLE
SUPPLY PIPING

**Fig. 9-34.   Using a compression union to join ⅜" copper tubing
from faucet to ⅜" flexible supply piping.**

Trough urinals, some of which are still in use, are prohibited by
most plumbing codes. Old installations may be allowed to remain
but when repair or replacement is needed, trough urinals must be
replaced with other types. Urinals are made of vitreous china and
are designed to be self-cleaning.

Stall Urinals

A stall urinal is shown in Fig. 9-35. Stall urinals, by the nature of
their design and the way they are used, leave much to be desired
in the way of sanitation. In addition, the trap, being placed under
the floor, is not easily accessible for rodding-out, and because users

Fig. 9-35. A stall urinal.

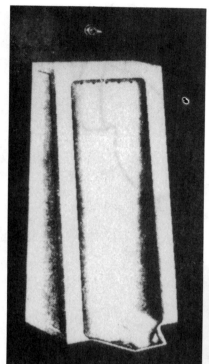

Fig. 9-36. A wall-hung vitreous china urinal. *(Courtesy Eljer Plumbingware)*

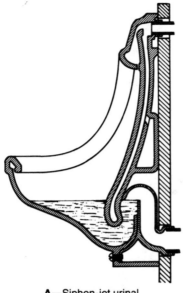

A—Siphon-jet urinal

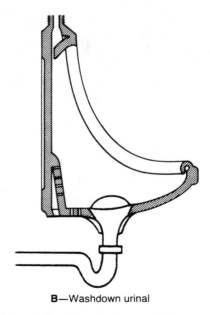

B—Washdown urinal

Fig. 9-37. Cut-away views of siphon-jet and washdown urinals.

do not flush the fixture with each use, mineral deposits build up in the trap, contributing to frequent stoppages.

Wall-hung Urinals

A typical wall-hung urinal is shown in Fig. 9-36. This fixture has an integral (built-in) trap, as shown in Fig. 9-37(A), and with liquid visible, tends to be flushed by each user if a hand-operated flushing mechanism is installed. Wall-hung wash-down urinals, Fig. 9-37(B), contain no visible liquid; the trap is located beneath the fixture. Because no liquid is visible, many users tend not to flush this fixture with each use and strong odors are present.

Pedestal urinals are made with an integral trap, the waste connects at the floor. A pedestal urinal is shown in Fig. 9-38.

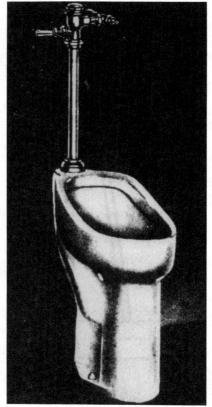

Fig. 9-38. A pedestal urinal.
(Courtesy Crane Plumbing)

Bidet

The *bidet* (pronounced be-day') is a relatively new fixture in this country; however, it has been in common use in the Latin American countries and in Europe for many years (Fig. 9-39). This fixture can be a valuable contribution to personal cleanliness and sensible living for every member of the family. The bidet is designed for cleanliness of the localized parts of the body, and it serves many useful purposes. Its use is a clean habit for men, women, and children. Its frequent application is advisable for comfort and health and in keeping with a mode of sanitary living.

Americans who have traveled either in Europe or in South America have come to accept the bidet as a standard bathroom fixture, and they have learned that it is a logical twin fixture to the toilet and a remarkable aid to personal cleanliness. Since more Americans have traveled abroad and become acquainted with the advantages of the bidet, more of these fixtures are being installed in new homes. Doctors who emphasize the basic importance of genitourinary cleanliness are often recommending the bidet and the washing practices made possible by the bidet to their patients of both sexes. In the near future, the bidet may be accepted as a necessary bathroom fixture.

The bidet is equipped with valves for both hot and cold water and with a pop-up type waste plug either for retaining the water or for draining it as desired. The inside walls of the bowl are washed by a flushing rim that uses the same basic principle of operation as the toilet bowl; however, the bidet is neither designed nor intended to carry away human waste material.

The fixture is also provided with an integral douche or jet, operated when desired by means of a transfer valve which directs a stream or column of water upward from the bottom section of the bowl (Fig. 9-40). This jet is formed by a cluster of small holes which are arranged to direct the water to a central point, thereby forming a solid stream of water.

The bidet is also used as a foot bath. Since the distance from the floor to the top of the rim is 14 to 15 in., it can be used very comfortably as a foot bath. Since the bidet is made from white vitreous china, there should be no hesitancy in using it as a foot bath. By means of the flushing rim, it can be kept clean and sanitary by merely rinsing with water or wiping with a damp cloth.

Fig. 9-39.   A vitreous china bidet. *(Courtesy Crane Plumbing)*

Physicians may advise the use of the bidet for individuals who cannot use the bathtub or shower because of ill health; it helps elderly people to bathe without exertion. The thermal effect and soothing action created by water under pressure is often recommended for irritations of the skin or following operations or injuries to the pelvic area.

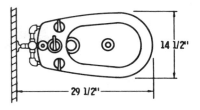

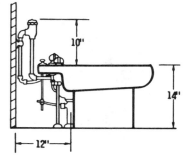

**Fig. 9-40.   A vitreous china bidet showing spray head and rough-
ing-in measurements.** *(Courtesy Crane Plumbing)*

Thermostatically Controlled Water Valves

The temperature of the water for the gang-type shower baths installed in many schools and factories can be controlled automatically.
One of these control devices is shown in Fig. 9-41. The temperature
of the water can be thermostatically controlled for hospital hydro-
therapy, film processing, plastic molding, and many industrial pro-
cesses. The control mixes the hot and cold water and delivers the
blended mixture at the desired thermostatically controlled temperature. A liquid-filled expansion-type thermostat, contained within
the assembly, maintains the selected water temperature even though
both hot- and cold-water supplies may fluctuate in temperature and
pressure.

In the control device shown (Fig. 9-41), the water flow from
the mixer is stopped automatically and immediately in event of a
failure in the supply of cold water. A concealed adjustment for raising
and lowering the temperature range is included. A single handle
selects the temperature of the water delivered, and it also shuts off
the discharge of water. The control device is constructed with only
one moving part and is accessible from the front. It is available in
two different temperature ranges—65°F to 110°F and 75°F to 175°F.

A dial-type thermometer (Fig. 9-43) is available either in rigid
or remote bulb-type instruments that are placed in a pipe line for
immersion in a liquid or for air exposure. The instrument is available
in a wide variety of temperature ranges. This type of thermometer
permits a visible check on the operating temperature of the water
for a gang-type shower bath.

Fig. 9-41.   A thermostatically controlled water valve. *(Courtesy The
Powers Regulator Co.)*

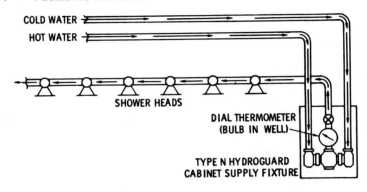

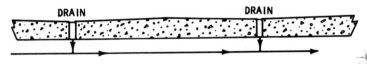

Fig. 9-42. A thermostatically controlled water valve installed in a cabinet. *(Courtesy The Powers Regulator Co.)*

Fig. 9-43. A dial thermometer permits a visible check on water temperature.

The control fixture and the dial-type thermometer can be enclosed in a cabinet, with only authorized personnel having access to the cabinet (Fig. 9-42). The dial-type thermometer permits a visible check on the operating temperature of the water for the shower bath at any time.

CHAPTER 10

Lead Work

Lead has virtually vanished from the plumbing scene in most areas of the country. Today, its main use is making lead bends for toilets. Nevertheless, knowing how to make a wiped joint—one in which lead or other sections are soldered together—is still required in many examinations for plumbing licenses. Hence, what follows is a detailed description of how to make a wiped joint. If you do have the occasion to work with lead, it will be well to observe the precautions for handling it that were outlined in Chapter 1 of this volume.

Judging the Solder

A requirement in lead work is the ability to judge the quality of solder. The plumber must know by its appearance when it contains the right proportions of lead and tin. To preserve these proportions, it is necessary to keep the solder from overheating because in overheating some of the tin will burn, thus destroying the correct proportions. The tin burns because its melting point is lower than that of the lead. The quality of the solder may be judged by pouring out a small quantity on a brick or stone and noting the color when it sets, and the number and size of bright spots on its surface. When the proportions are correct, there will appear on a test sample (almost the size of a half dollar) three or four small bright spots. The side of the solder next to the brick will be bright. Adding lead will reduce the size and number of spots; adding tin will increase them. Thoroughly stir the solder before pouring out a test sample. The rate of cooling affects the appearance of the test sample; if cooled too quickly, the solder will appear *finer* than it really is.

When the tin burns, it is indicated by the formation of dross on the surface, specks of which turn bright red and smoke. Too little tin in the solder will cause the solder to melt the lead pipe on which it is poured; it will burn the tinning of a brass ferrule or union and set free zinc from the brass, which will mix with the solder and render it unfit for joint wiping. The right heat of the solder is judged by the color or bloom on the surface of the molten solder, or by holding the ladle near the face. An easier test for the beginner is to stir with a wooden stick; when it is at the right temperature it will char the stick; if it is too hot, the stick will burn.

Proportioning the Solder

Wiping solder is composed of two parts lead and one part tin. In using wiping solder, the numerous heatings and occasional overheatings will result in loss of some of the tin content. It is necessary to add a little tin from time to time to restore the proper balance of lead and tin. Since tin is lighter than lead, it tends to float on top of the lead, and unless the wiping solder is stirred before a ladleful is removed, an excess amount of tin will be removed. Table 10-1 shows the properties of lead and tin.

The following method of making wiping solder is recommended: Melt down 20 lbs. of lead and 10 lbs. of tin, using a new or clean lead pot. When the lead and tin have melted, throw in 2 oz. of rosin, and stir well. When heated to 600° the wiping solder

Table 10-1   Properties of Lead and Tin

Ingredients	Melting Point	Specific Gravity	Weight	
			Per cu. in.	Per cu. ft.
Lead	620°F	11.07 to 11.44	.4106	709.7
Tin	449°F	7.297 to 7.409	.2652	458.3

is ready for use. A piece of newspaper submerged in the solder will ignite at 600°.

Preparing Joint for Wiping

It is important that the ends of the lead pipes to be joined are properly treated before wiping. The two essential requirements for a satisfactory flow of liquid through the pipe in service are as follows:

1. That the ends of the pipes to be joined properly fit, so that in pouring the solder, it will not run through the joint and form an obstruction.
2. That there are no sharp internal projections at the joint that would catch lint or any other foreign matter.

In addition, the formation given to the ends of the pipes should be such as to form a socket into which the solder will flow, thus making the joint stronger than if merely built up around the outer surfaces of the two pipes.

The operations to be performed in preparing the joint for wiping consist of:

1. Squaring.
2. Removing burrs.
3. Flaring the female end.
4. Rasping the outer edge.
5. Pointing the male end.
6. Soiling.
7. Marking.
8. Shaving.
9. Setting.

Various tools are used in performing these operations.

To secure a good fitting joint, so that when the solder is poured it will not run inside the pipes, the ends of the pipe must first be squared, as shown in Fig. 10-1. Cut the pipe as true as possible. The skilled workman will be able to judge when the end is square by eye, but the beginner should use a try square to test the trueness of the end. When the pipe is cut, especially if a wheel cutter is used (such as shown in Fig. 10-2), burrs will be formed on the inside and outside of the pipe. At this stage, the inside burr should be removed

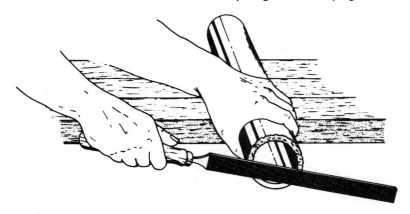

Fig. 10-1.   Preparing a joint for wiping by squaring the end.

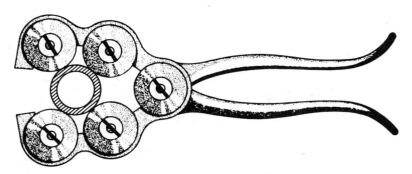

Fig. 10-2.   Wheel cutter for lead pipe.

by using a reamer, tap borer, or a shave hook, as shown in Fig. 10-3.

In the further preparation of the ends, the *female end* is flared or belled out with a *turn pin* as shown in Fig. 10-4. The pipe is flared so that the end is enlarged about a quarter of an inch. The result is shown in Fig. 10-5. After flaring, the outer burrs should be removed with a rasp, holding it in a plane parallel to the surface of the pipe as shown in Fig. 10-6. This is done to reduce the amount of solder necessary in wiping. The next step is to *point* the *male*

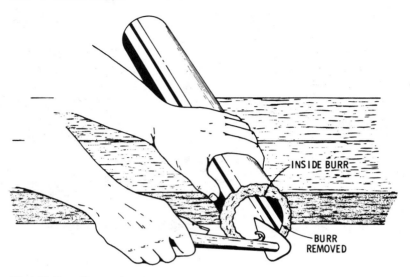

Fig. 10-3. Removing burrs from end of pipe.

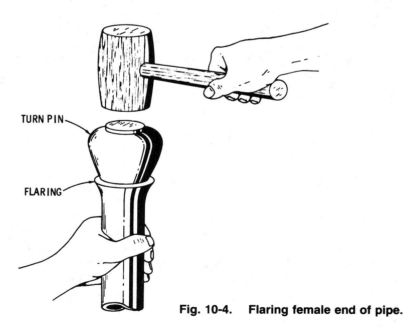

Fig. 10-4. Flaring female end of pipe.

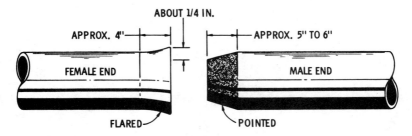

Fig. 10-5. Shape of female and male ends of pipe.

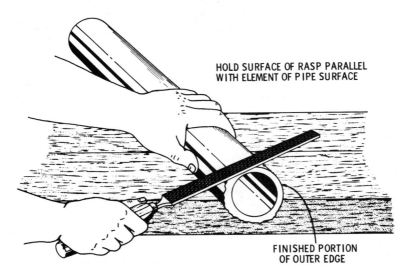

HOLD SURFACE OF RASP PARALLEL
WITH ELEMENT OF PIPE SURFACE

FINISHED PORTION
OF OUTER EDGE

Fig. 10-6. Rasping outer edge of female end of pipe.

end with a rasp, as shown in Fig. 10-7. The taper on this end should be somewhat longer than on the other end to permit sweating, which is desirable as it increases the strength of the joint. This is shown in the enlarged section in Fig. 10-8. In pointing, the fit of the two ends should be frequently tested until the fit shown in Fig. 10-8 is approximated. The ends are now ready for soiling.

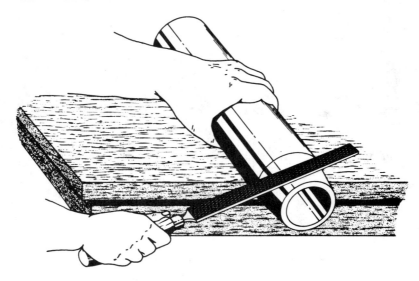

Fig. 10-7.   Pointing the male end of pipe.

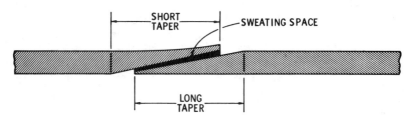

Fig. 10-8.   Enlarged section of the male and female pipe.

First, remove all grease or oil from the pipe by rubbing the surface with chalk, sand, or wire cloth, thus presenting a clean surface to which the soil will adhere. The soil is a composition of lamp black mixed with a little glue and water; it is painted around the pipe (as shown in Fig. 10-9) to prevent the adhesion of the melted solder except at its proper place, thus giving a neat and finished appearance. Ready-mixed plumbers' soil may also be obtained. In the absence of the prepared article, use old-fashioned shoe blacking; this, however, is not as satisfactory as regular soil.

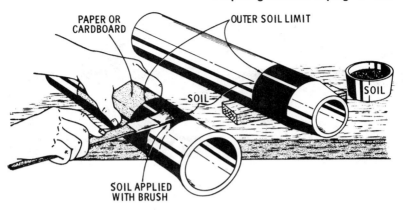

Fig. 10-9.   Applying the soil to pipe.

To make the soil, take ½ oz. of pulverized glue and dissolve it in water, and gradually add a pint of dry lamp black with water enough to bring it to the consistency of cream. Boil and stir until the glue is thoroughly incorporated with the black. This will have to be done slowly, and when it has progressed far enough, test it as follows: Paint a little of the soil on a piece of pipe, and when dry, rub it smartly with your finger. If it comes off easily, add more glue, but if it sticks and takes a slight polish, it is good. If it curls off when heat is applied, there is too much glue in it, or the pipe was not cleaned prior to applying.

The entire end of each pipe is painted, with the soil extending beyond the joint limit as shown in Fig. 10-9. For neatness, paint the outer soil limit (on both pipes) to the lines by wrapping a piece of paper or cardboard around the pipe with the edge at the desired outer soil limit. After the soil dries, the excess must be removed from the pipe end up to the inner soil limit, which governs the length of the joint or the distance along the pipe to which the solder will adhere. The pipe ends are now ready for *shaving*. This consists of removing the soil between the pipe end and the inner soil limit to order to obtain a clean bright surface to which the solder will adhere. Both the internal and external surfaces must be shaved so that all the surface that comes in contact with the solder will be bright; otherwise the solder will not adhere.

Immediately after shaving, apply a little tallow to the shaved

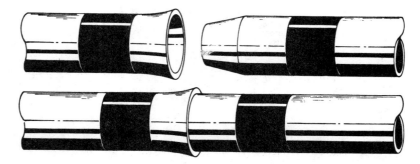

Fig. 10-10. Soil applied to pipe end for joint wiping.

surfaces to preserve them from the oxidizing action of the atmosphere, which would otherwise tarnish the surface and form a film to which the solder will not adhere. The pipes are now ready for the final preparatory operation of setting. They have the appearance shown in Fig. 10-10.

Setting the pipes or fixing them rigidly in position so that they will not move during the wiping operation often taxes the ingenuity of the workman. It is an easy job on the bench, but in a building, between beams, or in other cramped places, it is often very difficult to get proper support and leave room for manipulating the solder. In bench work, the pipe may be set either with blocks and string, or with clamps.

In setting (Fig. 10-11), the pipes are supported on four blocks. At intermediate points on both sides of the pipes, nails are driven. A string is attached to the end nail and a turn taken around the opposite nail drawing the string taut; it is carried to the next nail, and the operation repeated for each pair of nails.

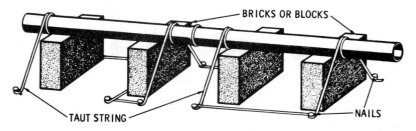

Fig. 10-11. Supporting lead pipe in preparing joint for wiping.

Table 10-2 Lengths of Wiped Joints

Pipe diameter (inches)	One-hand system		Two-hand system	
	Length of joint (inches)	Size of cloth (inches)	Length of joint (inches)	Size of cloth (inches)
½	2	3 × 3	2¼	3 × 4
¾	2	3 × 3	2⅜	3 × 4
1	2	3 × 3	2⅜	3 × 4
1¼ water	2	3 × 3	2½	3¼ × 4
1¼ waste	2	3 × 3	2⅜	3 × 4
1½ water	2	3 × 3	2½	3¼ × 4
1½ waste	2	3 × 3	2⅜	3 × 4
2 waste	2	3 × 3	2⅜	3¼ × 4
3 waste	2	3 × 3	2½	3¼ × 4
.4 waste	1¾	3 × 3, 6 × 6	2¾	3¼ × 4, 3¼ × 5
2 vertical	1¾	3 × 3	2	3 × 2½
3 vertical	1¾	3 × 3	2	3 × 2½
4 vertical	1¾	3 × 3	2	3 × 2½

Length of Joint

For guidance, Table 10-2 gives the length of joints for various-size pipes. The lengths specified in Table 10-2 represent the average American practice and will be found amply large for strength and durability, and the proportions give a pleasing appearance. The table also gives the size of wiping cloths.

Wiping the Joint

After the pipe ends have been prepared, as just described, they are ready for the final operation of wiping. The tools needed are the furnace pot (Fig. 10-12) and ladle (Fig. 10-13) for melting and dipping out the solder, and a wiping cloth. The information in Table 10-3 gives the amount of solder required for wiping joints of various sizes of pipe.

For joints up to 2 in. in diameter, a pot containing 10 lbs. of solder will ordinarily be large enough.

Fig. 10-12. A typical melting pot.

Fig. 10-13. A ladle for pouring solder.

There are three methods of wiping:

1. One-hand.
2. Two-hand.
3. Rolling.

On making a joint by the one-hand method, a quantity of solder is taken from the pot by means of the ladle, the solder being previously heated so hot that the hand cannot be held closer than two inches from the surface. The solder is poured lightly on the joint, the ladle being moved backward and forward, so that too much solder is not put in one place. The solder is also poured an inch or two on the soiling, to make the pipe the proper temperature. Naturally, the further the heat is run or taken along the pipe, the better the chance of making the joint. The operator keeps pouring, and

Size of Pipe (inches) ...	½	¾	1	1¼	1¼ water	1½ waste
Ounces of Solder	9	12	16	16	18	18

Size of Pipe (inches) ...	1½ water	2 waste	3 waste	4 waste	4 vertical
Ounces of Solder	20	20	24	34	28

with his left hand holds the cloth to catch the solder and to tin the lower side of the pipe, and also to keep the solder from dripping down. By the process of steady pouring, the solder now becomes soft and begins to feel shaped, firm, and bulky.

When in this shape and in a semifluid condition, the ladle is put down and, with the left hand, the operation of wiping (Fig. 10-14) is begun, working from the soiling toward the top of the bulb. If the solder cools rapidly, it is reheated to a plastic condition by a torch or a heated iron. When the joint is completed, it is cooled with a water spray so that the solder does not have time to alter its shape. The cloth used for wiping is a pad of moleskin or fustian about 4 in. square made from a piece of 9 × 12-in. material, folded six times and sewed to keep it from opening. The side next to the pipe is saturated with hot tallow when used. If the lead has been brought to the heat of the solder, and the latter properly manipulated and shaped while in a semifluid or plastic condition, the joint gradually assumes the finished egg-shaped appearance.

In wiping by the two-hand method, as soon as there is a sufficient body of solder around the pipe to retain the heat long enough for the wiping operation, drop the ladle and pick up a small cloth known as the auxiliary cloth. This is held in the right hand and the

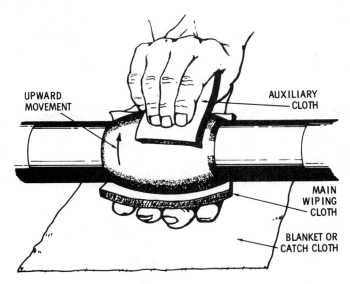

Fig. 10-14.   Wiping a horizontal joint, two-hand method.

Fig. 10-15.   Wiping a vertical joint.

wiping cloth in the left hand. The metal is brought to the top of the joint by a movement of both hands, as shown in Fig. 10-14. Hold the main cloth under the joint, and with the auxiliary cloth wipe off the surplus solder from each end. Roughly mold what is left on top to the shape of the joint, throwing all the hot solder into the wiping cloth. Stock this surplus solder to the bottom of the joint and roughly mold it to the proper shape. Drop the auxiliary cloth, and finish the joint to shape with the main cloth, using both hands.

In wiping a vertical joint, a small piece of cardboard is placed under the joint to catch excess solder, as shown in Fig. 10-15, forming a flange held in place around the pipe by twine.

Wiping a Branch Joint

Usually more skill is required in preparing and wiping a branch joint than a regular joint. The operations of preparing the joint for wiping are:

1. Boring.
2. Expanding.
3. Flaring out.
4. Removing burrs.
5. Soiling.
6. Shaving.
7. Setting.

First, the pipe from which a branch is to run is tapped with a tap borer, as in Fig. 10-16. In using a tap borer, do not insert it far enough for its point to come into contact with the opposite side of the pipe. For ½- to 1-in. pipe, bore a ⅝-in. hole. The operations that follow consist of flaring out (Fig. 10-17), removing burrs, soiling,

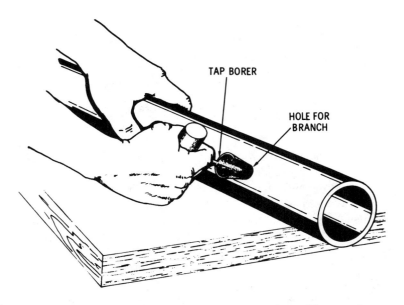

Fig. 10-16. Cutting a branch hole in a lead pipe with a tap borer.

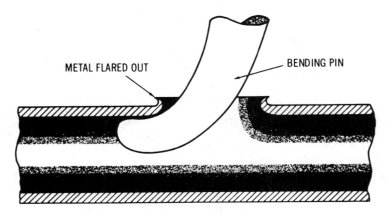

Fig. 10-17. Flaring a branch hole in lead pipe.

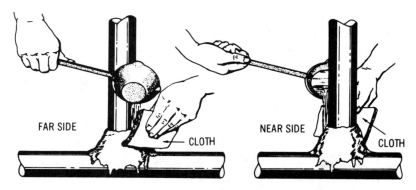

Fig. 10-18.   Pouring and wiping vertical branches.

marking off, shaving, and setting, which are performed in a way similar to those for plain or running joints.

In setting, the parts should be secured firmly in position with clamps, blocks, etc. It will be found easier to wipe the joint by setting up the branch in the vertical position. In wiping, pour on the far and near sides, as shown in Fig. 10-18, holding the cloth at an angle that will distribute the solder over the area to be covered. As the solder begins to flow, it is kept working up by manipulating the cloth. When sufficient solder has been poured to form the joint, the plumber first puts it roughly into shape with the cloth, followed by the wiping movements. The first wiping stroke encircles the

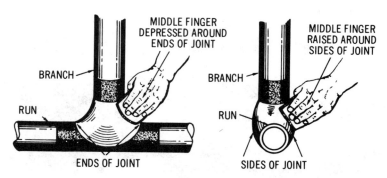

Fig. 10-19.   Movements in wiping vertical branches.

branch, the solder being shaped by depressing the middle finger as the cloth is being brought around the ends of the joint, gradually raising this finger as it comes on the sides parallel to the run. These operations are shown in Fig. 10-19. The stroke should begin on the near side, as far around as possible, so the operator can entirely encircle the branch with one stroke.

Index

AUDEL®

**Over a Century of Excellence
for the Professional
and
Vocational Trades and the Crafts**

Order now from your local bookstore
or use the convenient order form
at the back of this book.

AUDEL

These fully illustrated, up-to-date guides and manuals mean a better job done for mechanics, engineers, electricians, plumbers, carpenters, and all skilled workers.

CONTENTS

ELECTRICAL

HOUSE WIRING (Sixth Edition)
ROLAND E. PALMQUIST
5 1/2 x 8 1/4 Hardcover 256 pp. 150 Illus.
ISBN: 0-672-23404-1 $14.95
The rules and regulations of the National Electrical Code as they apply to residential wiring fully detailed with examples and illustrations.

PRACTICAL ELECTRICITY
(Fifth Edition)
ROBERT G. MIDDLETON;
revised by L. DONALD MEYERS
5 1/2 x 8 1/4 Hardcover 512 pp. 335 Illus.
ISBN: 0-02-584561-6 $19.95
The fundamentals of electricity for electrical workers, apprentices, and others requiring concise information about electric principles and their practical applications.

GUIDE TO THE 1987 NATIONAL ELECTRICAL CODE
ROLAND E. PALMQUIST
5 1/2 x 8 1/4 Hardcover 664 pp. 225 Illus.
ISBN: 0-02-594560-2 $22.50
The most authoritative guide available to interpreting the National Electrical Code for electricians, contractors, electrical inspectors, and homeowners. Examples and illustrations.

MATHEMATICS FOR ELECTRICIANS AND ELECTRONICS TECHNICIANS
REX MILLER
5 1/2 x 8 1/4 Hardcover 312 pp. 115 Illus.
ISBN: 0-8161-1700-4 $14.95
Mathematical concepts, formulas, and problem-solving techniques utilized on-the-job by electricians and those in electronics and related fields.

FRACTIONAL-HORSEPOWER ELECTRIC MOTORS
REX MILLER and
MARK RICHARD MILLER
5 1/2 x 8 1/4 Hardcover 436 pp. 285 Illus.
ISBN: 0-672-23410-6 $15.95
The installation, operation, maintenance, repair, and replacement of the small-to-moderate-size electric motors that power home appliances and industrial equipment.

ELECTRIC MOTORS (Fourth Edition)
EDWIN P. ANDERSON;
revised by REX MILLER
5 1/2 x 8 1/4 Hardcover 656 pp. 405 Illus.
ISBN: 0-672-23376-2 $14.95
Installation, maintenance, and repair of all types of electric motors.

HOME APPLIANCE SERVICING (Fourth Edition)
EDWIN P. ANDERSON;
revised by REX MILLER
5 1/2 x 8 1/4 Hardcover 640 pp. 345 Illus.
ISBN: 0-672-23379-7 $22.50
The essentials of testing, maintaining, and repairing all types of home appliances.

TELEVISION SERVICE MANUAL (Fifth Edition)
ROBERT G. MIDDLETON;
revised by JOSEPH G. BARRILE

5 1/2 x 8 1/4   Hardcover   512 pp.   395 Illus.
ISBN: 0-672-23395-9 $16.95

A guide to all aspects of television transmission and reception, including the operating principles of black and white and color receivers. Step-by-step maintenance and repair procedures.

ELECTRICAL COURSE FOR APPRENTICES AND JOURNEYMEN (Third Edition)
ROLAND E. PALMQUIST

5 1/2 x 8 1/4   Hardcover   478 pp.   290 Illus.
ISBN: 0-02-594550-5 $19.95

This practical course in electricity for those in formal training programs or learning on their own provides a thorough understanding of operational theory and its applications on the job.

QUESTIONS AND ANSWERS FOR ELECTRICIANS EXAMINATIONS (Ninth Edition)
ROLAND E. PALMQUIST

5 1/2 x 8 1/4   Hardcover   316 pp.   110 Illus.
ISBN: 0-02-594691-9 $18.95

Based on the 1987 National Electrical Code, this book reviews the subjects included in the various electricians examinations—apprentice, journeyman, and master. Question and Answer format.

MACHINE SHOP AND MECHANICAL TRADES

MACHINISTS LIBRARY
(Fourth Edition, 3 Vols.)
REX MILLER

5 1/2 x 8 1/4   Hardcover 1352 pp.   1120 Illus.
ISBN: 0-672-23380-0 $52.95

An indispensable three-volume reference set for machinists, tool and die makers, machine operators, metal workers, and those with home workshops. The principles and methods of the entire field are covered in an up-to-date text, photographs, diagrams, and tables.

Volume I: Basic Machine Shop
REX MILLER

5 1/2 x 8 1/4   Hardcover   392 pp.   375 Illus.
ISBN: 0-672-23381-9 $17.95

Volume II: Machine Shop
REX MILLER

5 1/2 x 8 1/4   Hardcover   528 pp.   445 Illus.
ISBN: 0-672-23382-7 $19.95

Volume III: Toolmakers Handy Book
REX MILLER

5 1/2 x 8 1/4   Hardcover   432 pp.   300 Illus.
ISBN: 0-672-23383-5 $14.95

MATHEMATICS FOR MECHANICAL TECHNICIANS AND TECHNOLOGISTS
JOHN D. BIES

5 1/2 x 8 1/4   Hardcover   342 pp.   190 Illus.
ISBN: 0-02-510620-1 $17.95

The mathematical concepts, formulas, and problem-solving techniques utilized on the job by engineers, technicians, and other workers in industrial and mechanical technology and related fields.

MILLWRIGHTS AND MECHANICS GUIDE
(Fourth Edition)
CARL A. NELSON

5 1/2 x 8 1/4   Hardcover   1,040 pp.   880 Illus.
ISBN: 0-02-588591-x $29.95

The most comprehensive and authoritative guide available for millwrights, mechanics, maintenance workers, riggers, shop workers, foremen, inspectors, and superintendents on plant installation, operation, and maintenance.

WELDERS GUIDE (Third Edition)
JAMES E. BRUMBAUGH

5 1/2 x 8 1/4   Hardcover   960 pp.   615 Illus.
ISBN: 0-672-23374-6 $23.95

The theory, operation, and maintenance of all welding machines. Covers gas welding equipment, supplies, and process; arc welding equipment, supplies, and process; TIG and MIG welding; and much more.

WELDERS/FITTERS GUIDE
JOHN P. STEWART

8 1/2 x 11   Paperback   160 pp.   195 Illus.
ISBN: 0-672-23325-8 $7.95

Step-by-step instruction for those training to become welders/fitters who have some knowledge of welding and the ability to read blueprints.

SHEET METAL WORK

JOHN D. BIES

5 1/2 x 8 1/4   Hardcover   456 pp.   215 Illus.
ISBN: 0-8161-1706-3 $19.95

An on-the-job guide for workers in the manufacturing and construction industries and for those with home workshops. All facets of sheet metal work detailed and illustrated by drawings, photographs, and tables.

POWER PLANT ENGINEERS GUIDE (Third Edition)

FRANK D. GRAHAM;
revised by CHARLIE BUFFINGTON

5 1/2 x 8 1/4   Hardcover   960 pp.   530 Illus.
ISBN: 0-672-23329-0 $27.50

This all-inclusive, one-volume guide is perfect for engineers, firemen, water tenders, oilers, operators of steam and diesel-power engines, and those applying for engineer's and firemen's licenses.

MECHANICAL TRADES POCKET MANUAL (Second Edition)

CARL A. NELSON

4 x 6   Paperback   364 pp.   255 Illus.
ISBN: 0-672-23378-9   10.95

A handbook for workers in the industrial and mechanical trades on methods, tools, equipment, and procedures. Pocket-sized for easy reference and fully illustrated.

PLUMBING

PLUMBERS AND PIPE FITTERS LIBRARY (Fourth Edition, 3 Vols.)

CHARLES N. McCONNELL

5 1/2 x 8 1/4   Hardcover   952 pp.   560 Illus.
ISBN: 0-02-582914-9 $68.45

This comprehensive three-volume set contains the most up-to-date information available for master plumbers, journeymen, apprentices, engineers, and those in the building trades. A detailed text and clear diagrams, photographs, and charts and tables treat all aspects of the plumbing, heating, and air conditioning trades.

Volume I: Materials, Tools, Roughing-In

CHARLES N. McCONNELL;
revised by TOM PHILBIN

5 1/2 x 8 1/4   Hardcover   304 pp.   240 Illus.
ISBN: 0-02-582911-4 $20.95

Volume II: Welding, Heating, Air Conditioning

CHARLES N. McCONNELL;
revised by TOM PHILBIN

5 1/2 x 8 1/4   Hardcover   384 pp.   220 Illus.
ISBN: 0-02-582912-2 $22.95

Volume III: Water Supply, Drainage, Calculations

CHARLES N. McCONNELL;
revised by TOM PHILBIN

5 1/2 x 8 1/4   Hardcover   264 pp.   100 Illus.
ISBN: 0-02-582913-0 $20.95

HOME PLUMBING HANDBOOK (Third Edition)

CHARLES N. McCONNELL

8 1/2 x 11   Paperback   200 pp.   100 Illus.
ISBN: 0-672-23413-0 $13.95

An up-to-date guide to home plumbing installation and repair.

THE PLUMBERS HANDBOOK (Seventh Edition)

JOSEPH P. ALMOND, SR.

4 x 6   Paperback   352 pp.   170 Illus.
ISBN: 0-672-23419-x $11.95

A handy sourcebook for plumbers, pipe fitters, and apprentices in both trades. It has a rugged binding suited for use on the job, and fits in the tool box or conveniently in the pocket.

QUESTIONS AND ANSWERS FOR PLUMBERS EXAMINATIONS (Second Edition)

JULES ORAVITZ

5 1/2 x 8 1/4   Paperback   256 pp.   145 Illus.
ISBN: 0-8161-1703-9 $9.95

A study guide for those preparing to take a licensing examination for apprentice, journeyman, or master plumber. Question and answer format.

HVAC

AIR CONDITIONING: HOME AND COMMERCIAL (Second Edition)

EDWIN P. ANDERSON;
revised by REX MILLER

5 1/2 x 8 1/4   Hardcover   528 pp.   180 Illus.
ISBN: 0-672-23397-5 $15.95

A guide to the construction, installation, operation, maintenance, and repair of home, commercial, and industrial air conditioning systems.

HEATING, VENTILATING, AND AIR CONDITIONING LIBRARY
(Second Edition, 3 Vols.)
JAMES E. BRUMBAUGH
5 1/2 x 8 1/4 Hardcover 1,840 pp. 1,275 Illus.
ISBN: 0-672-23388-6 $53.95
An authoritative three-volume reference library for those who install, operate, maintain, and repair HVAC equipment commercially, industrially, or at home.

Volume I: Heating Fundamentals, Furnaces, Boilers, Boiler Conversions
JAMES E. BRUMBAUGH
5 1/2 x 8 1/4 Hardcover 656 pp. 405 Illus.
ISBN: 0-672-23389-4 $17.95

Volume II: Oil, Gas and Coal Burners, Controls, Ducts, Piping, Valves
JAMES E. BRUMBAUGH
5 1/2 x 8 1/4 Hardcover 592 pp. 455 Illus.
ISBN: 0-672-23390-8 $17.95

Volume III: Radiant Heating, Water Heaters, Ventilation, Air Conditioning, Heat Pumps, Air Cleaners
JAMES E. BRUMBAUGH
5 1/2 x 8 1/4 Hardcover 592 pp. 415 Illus.
ISBN: 0-672-23391-6 $17.95

OIL BURNERS (Fourth Edition)
EDWIN M. FIELD
5 1/2 x 8 1/4 Hardcover 360 pp. 170 Illus.
ISBN: 0-672-23394-0 $16.95
An up-to-date sourcebook on the construction, installation, operation, testing, servicing, and repair of all types of oil burners, both industrial and domestic.

REFRIGERATION: HOME AND COMMERCIAL (Second Edition)
EDWIN P. ANDERSON;
revised by REX MILLER
5 1/2 x 8 1/4 Hardcover 768 pp. 285 Illus.
ISBN: 0-672-23396-7 $19.95
A reference for technicians, plant engineers, and the home owner on the installation, operation, servicing, and repair of everything from single refrigeration units to commercial and industrial systems.

PNEUMATICS AND
HYDRAULICS

HYDRAULICS FOR OFF-THE-ROAD EQUIPMENT (Second Edition)
HARRY L. STEWART;
revised by TOM PHILBIN
5 1/2 x 8 1/4 Hardcover 256 pp. 175 Illus.
ISBN: 0-8161-1701-2 $13.95

This complete reference manual on heavy equipment covers hydraulic pumps, accumulators, and motors; force components; hydraulic control components; filters and filtration, lines and fittings, and fluids; hydrostatic transmissions; maintenance; and troubleshooting.

PNEUMATICS AND HYDRAULICS (Fourth Edition)
HARRY L. STEWART;
revised by TOM STEWART
5 1/2 x 8 1/4 Hardcover 512 pp. 315 Illus.
ISBN: 0-672-23412-2 $19.95
The principles and applications of fluid power. Covers pressure, work, and power; general features of machines; hydraulic and pneumatic symbols; pressure boosters; air compressors and accessories; and much more.

PUMPS (Fourth Edition)
HARRY STEWART;
revised by TOM PHILBIN
5 1/2 x 8 1/4 Hardcover 508 pp. 360 Illus.
ISBN: 0-672-23400-9 $17.95
The principles and day-to-day operation of pumps, pump controls, and hydraulics are thoroughly detailed and illustrated.

CARPENTRY AND
CONSTRUCTION

CARPENTERS AND BUILDERS LIBRARY (Fifth Edition, 4 Vols.)
JOHN E. BALL; revised by TOM PHILBIN
5 1/2 x 8 1/4 Hardcover 1,224 pp. 1,010 Illus.
ISBN: 0-672-23369-x $43.95

Also available as a boxed set at no extra cost:
ISBN: 0-02-506450-9 $43.95

This comprehensive four-volume library has set the professional standard for decades for carpenters, joiners, and woodworkers.

Volume I: Tools, Steel Square, Joinery
JOHN E. BALL; revised by TOM PHILBIN
5 1/2 x 8 1/4 Hardcover 384 pp. 345 Illus.
ISBN: 0-672-23365-7 $10.95

Volume II: Builders Math, Plans, Specifications
JOHN E. BALL; revised by TOM PHILBIN
5 1/2 x 8 1/4 Hardcover 304 pp. 205 Illus.
ISBN: 0-672-23366-5 $10.95

Volume III: Layouts, Foundations, Framing
JOHN E. BALL; revised by TOM PHILBIN
5 1/2 x 8 1/4 Hardcover 272 pp. 215 Illus.
ISBN: 0-672-23367-3 $10.95

Volume IV: Millwork, Power Tools, Painting
JOHN E. BALL; revised by TOM PHILBIN
5 1/2 x 8 1/4   Hardcover   344 pp.   245 Illus.
ISBN: 0-672-23368-1 $10.95

COMPLETE BUILDING CONSTRUCTION (Second Edition)
JOHN PHELPS; revised by TOM PHILBIN
5 1/2 x 8 1/4   Hardcover   744 pp.   645 Illus.
ISBN: 0-672-23377-0 $22.50
Constructing a frame or brick building from the footings to the ridge. Whether the building project is a tool shed, garage, or a complete home, this single fully illustrated volume provides all the necessary information.

COMPLETE ROOFING HANDBOOK
JAMES E. BRUMBAUGH
5 1/2 x 8 1/4   Hardcover   536 pp.   510 Illus.
ISBN: 0-02-517850-4 $29.95
Covers types of roofs; roofing and reroofing; roof and attic insulation and ventilation; skylights and roof openings; dormer construction; roof flashing details; and much more.

COMPLETE SIDING HANDBOOK
JAMES E. BRUMBAUGH
5 1/2 x 8 1/4   Hardcover   512 pp.   450 Illus.
ISBN: 0-02-517880-6 $24.95
This companion volume to the *Complete Roofing Handbook* includes comprehensive step-by-step instructions and accompanying line drawings on every aspect of siding a building.

MASONS AND BUILDERS LIBRARY (Second Edition, 2 Vols.)
LOUIS M. DEZETTEL;
revised by TOM PHILBIN
5 1/2 x 8 1/4   Hardcover   688 pp.   500 Illus.
ISBN: 0-672-23401-7 $27.95
This two-volume set provides practical instruction in bricklaying and masonry. Covers brick; mortar; tools; bonding; corners, openings, and arches; chimneys and fireplaces; structural clay tile and glass block; brick walls; and much more.

Volume I: Concrete, Block, Tile, Terrazzo
LOUIS M. DEZETTEL;
revised by TOM PHILBIN
5 1/2 x 8 1/4   Hardcover   304 pp.   190 Illus.
ISBN: 0-672-23402-5 $13.95

Volume 2: Bricklaying, Plastering, Rock Masonry, Clay Tile
LOUIS M. DEZETTEL;
revised by TOM PHILBIN
5 1/2 x 8 1/4   Hardcover   384 pp.   310 Illus.
ISBN: 0-672-23403-3 $13.95

WOODWORKING

WOOD FURNITURE: FINISHING, REFINISHING, REPAIRING (Second Edition)
JAMES E. BRUMBAUGH
5 1/2 x 8 1/4   Hardcover   352 pp.   185 Illus.
ISBN: 0-672-23409-2 $12.95
A fully illustrated guide to repairing furniture and finishing and refinishing wood surfaces. Covers tools and supplies; types of wood; veneering; inlaying; repairing, restoring, and stripping; wood preparation; and much more.

WOODWORKING AND CABINETMAKING
F. RICHARD BOLLER
5 1/2 x 8 1/4   Hardcover   360 pp.   455 Illus.
ISBN: 0-02-512800-0 $18.95
Essential information on all aspects of working with wood. Step-by-step procedures for woodworking projects are accompanied by detailed drawings and photographs.

MAINTENANCE AND REPAIR

BUILDING MAINTENANCE (Second Edition)
JULES ORAVETZ
5 1/2 x 8 1/4   Hardcover   384 pp.   210 Illus.
ISBN: 0-672-23278-2 $11.95
Professional maintenance procedures used in office, educational, and commercial buildings. Covers painting and decorating; plumbing and pipe fitting; concrete and masonry; and much more.

GARDENING, LANDSCAPING AND GROUNDS MAINTENANCE (Third Edition)
JULES ORAVETZ
5 1/2 x 8 1/4   Hardcover   424 pp.   340 Illus.
ISBN: 0-672-23417-3 $15.95
Maintaining lawns and gardens as well as industrial, municipal, and estate grounds.

HOME MAINTENANCE AND REPAIR: WALLS, CEILINGS AND FLOORS
GARY D. BRANSON

8 1/2 x 11   Paperback   80 pp.   80 Illus.
ISBN: 0-672-23281-2 $6.95

The do-it-yourselfer's guide to interior remodeling with professional results.

PAINTING AND DECORATING
REX MILLER and GLEN E. BAKER

5 1/2 x 8 1/4   Hardcover   464 pp.   325 Illus.
ISBN: 0-672-23405-x $18.95

A practical guide for painters, decorators, and homeowners to the most up-to-date materials and techniques in the field.

TREE CARE (Second Edition)
JOHN M. HALLER

8 1/2 x 11   Paperback   224 pp.   305 Illus.
ISBN: 0-02-062870-6 $9.95

The standard in the field. A comprehensive guide for growers, nursery owners, foresters, landscapers, and homeowners to planting, nurturing and protecting trees.

UPHOLSTERING (Updated)
JAMES E. BRUMBAUGH

5 1/2 x 8 1/4   Hardcover   400 pp.   380 Illus.
ISBN: 0-672-23372-x $15.95

The essentials of upholstering fully explained and illustrated for the professional, the apprentice, and the hobbyist.

AUTOMOTIVE AND ENGINES

DIESEL ENGINE MANUAL
(Fourth Edition)
PERRY O. BLACK;
revised by WILLIAM E. SCAHILL

5 1/2 x 8 1/4   Hardcover   512 pp.   255 Illus.
ISBN: 0-672-23371-1 $15.95

The principles, design, operation, and maintenance of today's diesel engines. All aspects of typical two- and four-cycle engines are thoroughly explained and illustrated by photographs, line drawings, and charts and tables.

GAS ENGINE MANUAL
(Third Edition)
EDWIN P. ANDERSON;
revised by CHARLES G. FACKLAM

5 1/2 x 8 1/4   Hardcover   424 pp.   225 Illus.
ISBN: 0-8161-1707-1 $12.95

How to operate, maintain, and repair gas engines of all types and sizes. All engine parts and step-by-step procedures are illustrated by photographs, diagrams, and troubleshooting charts.

SMALL GASOLINE ENGINES
REX MILLER and MARK RICHARD MILLER

5 1/2 x 8 1/4   Hardcover   640 pp.   525 Illus.
ISBN: 0-672-23414-9 $16.95

Practical information for those who repair, maintain, and overhaul two- and four-cycle engines—including lawn mowers, edgers, grass sweepers, snowblowers, emergency electrical generators, outboard motors, and other equipment with engines of up to ten horsepower.

TRUCK GUIDE LIBRARY (3 Vols.)
JAMES E. BRUMBAUGH

5 1/2 x 8 1/4   2,144 pp.   1,715 Illus.
ISBN: 0-672-23392-4 $45.95

This three-volume set provides the most comprehensive, profusely illustrated collection of information available on truck operation and maintenance.

Volume 1: Engines
JAMES E. BRUMBAUGH

5 1/2 x 8 1/4   Hardcover   416 pp.   290 Illus.
ISBN: 0-672-23356-8 $16.95

Volume 2: Engine Auxiliary Systems
JAMES E. BRUMBAUGH

5 1/2 x 8 1/4   Hardcover   704 pp.   520 Illus.
ISBN: 0-672-23357-6 $16.95

Volume 3: Transmissions, Steering, and Brakes
JAMES E. BRUMBAUGH

5 1/2 x 8 1/4   Hardcover   1,024 pp.   905 Illus.
ISBN: 0-672-23406-8 $16.95

DRAFTING

INDUSTRIAL DRAFTING
JOHN D. BIES

5 1/2 x 8 1/4   Hardcover   544 pp.   Illus.
ISBN: 0-02-510610-4 $24.95

Professional-level introductory guide for practicing drafters, engineers, managers, and technical workers in all industries who use or prepare working drawings.

ANSWERS ON BLUEPRINT READING (Fourth Edition)

ROLAND PALMQUIST;
revised by THOMAS J. MORRISEY

5 1/2 x 8 1/4   Hardcover   320 pp.   275 Illus.
ISBN: 0-8161-1704-7 $12.95

Understanding blueprints of machines and tools, electrical systems, and architecture. Question and answer format.

HOBBIES

COMPLETE COURSE IN STAINED GLASS

PEPE MENDEZ

8 1/2 x 11   Paperback   80 pp.   50 Illus.
ISBN: 0-672-23287-1 $8.95

The tools, materials, and techniques of the art of working with stained glass.